高等教育"十三五"部委级规划教材

# 机械工程英语阅读教程

## （第三版）

| 主　　编 | 周益军 | | |
|---|---|---|---|
| 副 主 编 | 池寅生 | 何慧娟 | 单　敏 |
| 其他参编者 | 张建宏 | 肖淑梅 | 卫玉芬　孔纪兰 |
| | 刘　敏 | 徐小青 | 赵　翔　朱丹凤 |
| | 南丽霞 | 张承阳 | 张　翔 |
| 主　　审 | 孟咸智 | | |

东华大学 出版社

·上海·

图书在版编目(CIP)数据

机械工程英语阅读教程 / 周益军 主编. —3 版—上海:东华大学出版社,2019.4

ISBN 978-7-5669-1554-2

Ⅰ. ①机… Ⅱ. ①周… Ⅲ. ①机械工程—英语—阅读教学—高等职业教育—教材 Ⅳ. ①TH

中国版本图书馆 CIP 数据核字(2019)第 054015 号

责任编辑　曹晓虹
封面设计　刘　洋

## 机械工程英语阅读教程(第三版)

周益军　主编

出版发行　东华大学出版社
（上海市延安西路 1882 号　邮政编码:200051）
联系电话　编辑部　021-62379902
　　　　　营销中心　021-62193056　62373056
出版社网址　http://dhupress.dhu.edu.cn
天猫旗舰店　http://dhdx.tmall.com
印　　刷　江苏省南通市印刷总厂有限公司
开　　本　710mm×1000mm　1/16
印　　张　15
字　　数　679 千
版　　次　2019 年 4 月第 3 版
印　　次　2019 年 4 月第 1 次印刷

ISBN 978-7-5669-1554-2　　　　定价:37.90 元

# 编写说明

随着中国加入世贸组织,中国经济快速发展,经济全球化等,这些大背景给学生带来了更广阔的就业平台。但同时,对学生的英语水平也有了更高的要求。熟练的专业技术加上精良的专业英语知识无疑就是高技能、紧缺人才所必备的。因此,学好专业英语显得尤为重要。

本书是按照《高职高专院校机械工程类专业英语教学大纲》所编写的。编者在多年教学实践经验的基础上,力求按照行业培养的宽口径,使专业英语教材具有良好的通用性。并根据高职高专教育的应用性特征,使专业英语具有较强的实用性和针对性。

全书由材料科学、材料成形、机械制造、汽车、模具、机器人、CAD/CAM等七部分组成,共分成七个单元。每个单元均由课文、词汇、注释、相关练习和参考译文等五个部分组成。内容以材料工程、机械设计、机械制造、机电一体化、汽车制造与维修、模具设计与制造、数控技术、计算机辅助设计与制造等专业技术及其最新发展讯息为主,对所选的阅读材料中出现的语言重点和难点做了详细的注释。选材精炼,课文后配有生词短语表、注释和相应的练习,促进学生"学必思考,学练结合"。书后附有参考译文和练习答案,便于学生理解和核查自己学习与掌握的内容。

本书可作为高职高专机械设计与制造类、机电控制等专业的教材,也可供工商管理专业、经济类专业和英语专业学生、技术人员学习参考。建议教师根据各专业的学生情况,可不受教材编排顺序的限制,进行适当筛选。对老师没有选用的单元,学生可根据自己的兴趣和需要自学其中的部分内容,以拓宽专业英语的知识面。

本书由江苏省扬州职业大学周益军博士(副教授)任主编;南通纺织职业技术学院单敏老师任第一副主编;扬州职业大学池寅生硕士(讲师)任第二副主编。参编人员还有扬州职业大学肖淑梅博士、卫玉芬博士、孔纪兰博士、朱丹凤硕士、徐小青硕士、刘敏硕士、赵翔硕士、南丽霞硕士、张承阳硕士以及张翔博士等。

具体编写分工为：第一单元由孔纪兰编写；第二单元由朱丹凤编写；第三单元由周益军编写；第四单元，肖淑梅；第五单元，池寅生；第六单元，卫玉芬；第七单元，徐小青。单敏同志对本教材的词汇和练习进行了认真设计与编写。

本书的编写工作得到了扬州职业大学领导的高度重视与支持。孟咸智副教授对本教材的编写提出了宝贵意见，在此表示衷心的感谢。由于编者的水平和经验有限，书中难免有缺陷和不足之处，恳请广大读者批评指正。

编　者<br>2009 年 4 月

# 修订说明

本次修订以原教材为蓝本,保留了绝大多数内容。根据使用原教材的部分师生的反馈意见,编者对教材的部分内容进行了删减。删除了内容偏向阐述抽象理论为主的章节;保留了内容以介绍技术技能知识为主的部分。为了适应现代制造技术发展的需求,增补了3D打印技术和互联网+及中国制造2025等内容。在本次修订过程中,还增加了专业英语语法及相关词汇(构词法),以便学生学习。

3D打印技术由池寅生老师编写;互联网+及中国制造2025由张建宏老师编写;专业英语语法及相关词汇(构词法)部分由何慧娟老师编写。使用原教材的部分师生的反馈意见主要由何慧娟老师收集。

整个修订方案由周益军确定。在修订过程中,第一、二、三单元修订工作以何慧娟老师为主;第四、五、六、七单元修订工作以池寅生老师为主;第八单元审核校对工作以周益军老师为主。

本书修订后,由江苏省扬州职业大学周益军博士(教授)任主编;扬州职业大学池寅生硕士(讲师)、何慧娟硕士(讲师)、南通纺织职业技术学院单敏副教授任副主编。

编 者
2019年1月

# Contents

**Foreword**

**Unit 1**

| | | |
|---|---|---|
| **Lesson 1** | The History of Metallurgy | *1* |
| **Lesson 2** | Metallurgy | *5* |
| **Lesson 3** | Metal | *14* |
| **Lesson 4** | Materials in Industry | *20* |

**Unit 2**

| | | |
|---|---|---|
| **Lesson 1** | Casting | *28* |
| **Lesson 2** | Forging | *41* |
| **Lesson 3** | Welding | *50* |

**Unit 3**

| | | |
|---|---|---|
| **Lesson 1** | Lathe (metal) | *59* |
| **Lesson 2** | Milling Machine | *72* |
| **Lesson 3** | Types of Milling Cutter | *83* |

**Unit 4**

| | | |
|---|---|---|
| **Lesson 1** | Engine Operating Principles | *92* |
| **Lesson 2** | Engine Construction | *99* |

## Unit 5

| Lesson 1 | Forming of Sheet Metals | *106* |
| Lesson 2 | Metal Stamping Process and Die Design | *111* |
| Lesson 3 | Plastics and Injection Molding | *122* |
| Lesson 4 | Design of Injection Mold | *129* |

## Unit 6

| Lesson 1 | Food Robotics | *139* |
| Lesson 2 | Robotic Inspection | *152* |

## Unit 7

| Lesson 1 | Computer-Aided Design (CAD) | *165* |
| Lesson 2 | Computer Aided Manufacturing (CAM) | *171* |
| Lesson 3 | Computer Numerical Control | *180* |

## Unit 8

| Lesson 1 | 3D Printing Technology | *187* |
| Lesson 2 | "Internet +" & "Chinese wisdom made" | *197* |

## Answer to Exercises          *204*
## Vocabulary                   *216*

# Unit 1

## Lesson 1 The History of Metallurgy

*1* The earliest recorded metal employed by humans appears to be gold which can be found free or "native". Small amounts of natural gold has been found in Spanish caves used during the late Paleolithic period around 40,000 BC.

*2* Silver, copper, tin and meteoric iron can also be found native, allowing a limited amount of metalworking in early cultures. Egyptian weapons made from meteoric iron occurred about 3000 B. C.. However, by learning to get copper and tin by heating rocks and combining copper and tin to make an alloy called bronze, the technology of metallurgy began about 3500 B. C. in the Bronze Age.

*3* The extraction of iron from its ore into a workable metal is much more difficult. It appears to have been invented by the Hittites in about 1400 B. C., beginning the Iron Age. The secret of extracting and working iron was a key factor in the success of the Philistines.

Fig. 1.1 - 1  **Gold headband from Thebes 750-700 BC**

*4* Historical developments in ferrous metallurgy can be found in a wide variety of past cultures and civilizations. This includes the ancient and medieval kingdoms and empires of the Middle East and Near East, ancient Egypt and Anatolia (Turkey), Carthage, the Greeks and Romans of ancient Europe, medieval Europe, ancient and medieval China, ancient and medieval India, ancient and medieval Japan, etc. Of interest to note is that many applications, practices, and devices associated or involved in metallurgy were possibly established in ancient China before Europeans mastered these crafts (such as the innovation of the blast furnace, cast iron, steel, hydraulic-powered trip hammers, etc.).

Fig. 1.1 - 2  **Georg Agricola, author of De re metallica, an important early book on metal extraction**

*5* A 16th century book by Georg Agricola called *De re metallica* describes the highly developed and complex processes of mining metal ores, metal extraction and metallurgy of the time. Agricola has been described as the "father of metallurgy".

## ❖ New Words and Phrases

| | |
|---|---|
| metallurgy [meˈtælədʒi] n. | 冶金；冶金学；冶金术 |
| intermetallic [ˌintə(ː)miˈtælik] adj. | 金属间化合的 |
| compound [ˈkɔmpaund] n. | 混合物，化合物 |
| intermetallic compounds | 金属间化合物 |
| alloy [ˈælɔi] n. | 合金 |
| craft [krɑːft] n. | 工艺 |
| metalworking [ˈmetəlˌwɜːkiŋ] n. | 金属加工 |
| paleolithic [ˌpæliəuˈliθik] adj. | 旧石器时代的 |
| paleolithic period | 旧石器时代 |
| meteoric [ˌmiːtiˈɔrik] adj. | 流星的，昙花一现的 |
| meteoric iron | 陨铁 |
| bronze [brɔnz] n. | 青铜 |
| Bronze Age | 青铜器时代，青铜时代，铜器时代 |
| ferrous [ˈferəs] a. | 铁的，含铁的 |
| medieval [ˌmediˈiːvəl] a. | 中古的，中世纪的 |
| blast [blɑːst] n. | 爆破，冲击波 |
| furnace [ˈfəːnis] n. | 炉子，熔炉 |
| blast furnace | 鼓风炉，高炉 |
| hydraulic [haiˈdrɔːlik] a. | 水的，液压的 |
| trip hammer | 杵锤 |
| mining [ˈmainiŋ] n. | 采矿(业) |

## ❖ Notes

1. The earliest recorded metal employed by humans appears to be gold which can be found free or "native".
   最早记载的人类应用的金属,看起来是无偿获得的或"天然的"黄金。

**语法补充:which 引导定语从句**

一、本语法在注释一中的应用

　　which can be found free or "native"是定语从句,其中引导词 which 指代 gold。

二、对本语法的详细阐述

# Lesson 1　The History of Metallurgy

（一）which 可以引导限制性定语从句和非限制性定语从句。
a）限制性定语从句：与主句无逗号隔开，起修饰限制作用。
e. g. I like the coat which is very beautiful.
　　我喜欢这件非常漂亮的外套。
b）非限制性定语从句：与主句有逗号隔开，起补充说明作用。
e. g. This is a book, which I borrowed from my friend.
　　这是我从朋友那儿借来的一本书。
（二）which 引导定语从句时，可以指代前面一个词，也可以指代前面一句话。
a）指代前面一个词
e. g. He was reading a book, which was about war. （which 指代 a book）
　　他正在读一本关于战争的书。
b）指代前面一句话
e. g. He set free the birds happily, which was a celebration for his success. （which 指代 "He set free the birds happily"）
　　他开心地把鸟放了，这是对他成功的一种庆祝。
三、过去分词也可以起定语从句的作用，此时相当于省略了定语从句引导词 which 和 be 动词。如注释三中的 associated 和 involved。
e. g. This is a letter written （＝which is written） in German.
　　这是封用德语写的信。

2. However, by learning to get copper and tin by heating rocks and combining copper and tin to make an alloy called bronze, the technology of metallurgy began about 3500 B. C. with the Bronze Age.
然而，研究通过熔化矿石来制造铜和锡及通过熔化铜和锡来制造铜合金的冶金技术出现于大约公元前 3500 年的青铜时代。
3. Of interest to note is that many applications, practices, and devices associated or involved in metallurgy were possibly established in ancient China before Europeans mastered these crafts （such as the innovation of the blast furnace, cast iron, steel, hydraulic-powered trip hammers, etc.）.
值得注意的是，与冶金相关的许多应用、实践和设备，在中国古代出现的时间较欧洲早（如高炉，生铁，钢，水力杵锤等的发明）。

## Check your understanding

Ⅰ. Give brief answers to the following questions.
　　1. Who is called the "father of metallurgy"?

Ⅱ. Match the items listed in the following two columns.

| | |
|---|---|
| trip hammer | 鼓风炉，高炉 |
| compound | 金属加工 |
| blast furnace | 杵锤 |
| cast iron | 陨铁 |
| meteoric iron | 冶金；冶金学 |
| metalworking | 生铁 |
| metallurgy | 混合物 |

# 冶金学的历史

最早记载的人类应用的金属，看起来是无偿获得的或"天然的"黄金。在西班牙洞穴中发现了少量的天然黄金，这些洞穴的使用时间是在旧石器时代晚期(公元前40 000年左右)。

早期记载中也有天然的(含有限加工过的)银，铜，锡和陨铁等。用陨铁制造的埃及武器出现在公元前3 000年左右。然而，研究通过熔化矿石来制造铜和锡及通过熔化铜和锡来制造铜合金的冶金技术出现于大约公元前3 500年的青铜时代。

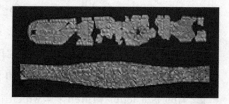

图1.1-1 公元前750-公元前700年底比斯的金头饰

从矿石中提取金属铁要困难得多。大约在公元前1 400年，赫梯人发明了炼铁术，进入铁器时代。非利士人成功的一个重要因素是拥有提炼和加工铁的秘密技术。

各种各样的历史文化和文明中都有钢铁冶金史。这包括古代和中世纪的中东和近东的王国和帝国、古埃及和土耳其，迦太基，希腊和古代欧洲罗马，中世纪的欧洲，中国古代和中世纪，古代和中世纪印度，古代和中世纪的日本等。值得注意的是，与冶金相关的许多应用、实践和设备，在中国古代出现的时间较欧洲早(如高炉，生铁，钢，水力杵锤等的发明)。

图1.1-2 格奥尔格·阿格里科拉,《论矿冶》一书的作者，早期重要的金属冶炼的书

16世纪由格奥尔格·阿格里科拉编著的《论矿冶》一书，描述了当时高度成熟和复杂的金属矿石的开采、金属提取和冶炼过程。阿格里科拉被称为"冶金之父"。

# Lesson 2　Metallurgy

*1*　Metallurgy is a domain of materials science that studies the physical and chemical behavior of metallic elements, their intermetallic compounds, and their mixtures, which are called alloys. It is also the technology of metals: the way in which science is applied to their practical use. Metallurgy is commonly used in the craft of metalworking.

## 1. Extractive metallurgy

*2*　Extractive metallurgy is the practice of removing valuable metals from an ore and refining the extracted raw metals into a purer form. In order to convert a metal oxide or sulfide to a purer metal, the ore must be refined either physically, chemically, or electrolytically.

**Fig. 1.2 - 1　Furnace bellows operated by waterwheels**

*3*　The picture above is an illustration of furnace bellows operated by waterwheels, from the *Nong Shu*, by Wang Zhen, 1313 AD, during the Chinese Yuan Dynasty.

*4*　Extractive metallurgists are interested in three primary streams: feed, concentrate (valuable metal oxide/sulfide), and tailings (waste). After mining, large pieces of the ore feed are broken through crushing and/or grinding in order to obtain particles small enough where each particle is either mostly valuable or mostly waste. Concentrating the particles of value in a form supporting separation enables the desired metal to be removed from waste products.

*5*　Mining may not be necessary if the ore body and physical environment are conducive to leaching. Leaching dissolves minerals in an ore body and results in an enriched solution. The solution is collected and processed to extract valuable metals.

*6*　Ore bodies often contain more than one valuable metal. Tailings of a previous process may be used as a feed in another process to extract a secondary product from the original ore. Additionally, a concentrate may contain more than one valuable metal. That concentrate would then be processed to separate the valuable metals into individual constituents.

## 2. Production engineering of metals

*7* In production engineering, metallurgy is concerned with the production of metallic components for use in consumer or engineering products. This involves the production of alloys, the shaping, the heat treatment and the surface treatment of the product. The task of the metallurgist is to achieve balance between material properties such as cost, weight, strength, toughness, hardness, corrosion and fatigue resistance, and performance in temperature extremes. To achieve this goal, the operating environment must be carefully considered. In a saltwater environment, ferrous metals and some aluminum alloys corrode quickly. Metals exposed to cold or cryogenic conditions may endure a ductile to brittle transition and lose their toughness, becoming more brittle and prone to cracking. Metals under continual cyclic loading can suffer from metal fatigue. Metals under constant stress at elevated temperatures can creep.

### 2.1 Metal working processes

*8* Metals are shaped by processes such as casting, forging, flow forming, rolling, extrusion, sintering, metalworking, machining and fabrication. With casting, molten metal is poured into a shaped mould. With forging, a red-hot billet is hammered into shape. With rolling, a billet is passed through successively narrower rollers to create a sheet. With extrusion, a hot and malleable metal is forced under pressure through a die, which shapes it before it cools. With sintering, a powdered metal is compressed into a die at high temperature. With machining, lathes, milling machines, and drills cut the cold metal to shape. With fabrication, sheets of metal are cut with guillotines or gas cutters and bent into shape.

*9* "Cold working" processes, where the product's shape is altered by rolling, fabrication or other processes while the product is cold, can increase the strength of the product by a process called work hardening. Work hardening creates microscopic defects in the metal, which resist further changes of shape.

*10* Various forms of casting exist in industry and academia. These include sand casting, investment casting (also called the "lost wax process"), die casting and continuous casting.

### 2.2 Joining

#### 2.2.1 Welding

*11* Welding is a technique for joining metal components cohesively by melting the base material, making the parts into a single piece. A filler material of similar composition (welding rod) may also be melted into the joint.

## 2.2.2 Brazing

*12* Brazing is a technique for joining metals adhesively at a temperature below their melting points. A filler with a melting point below that of the base metal is used, and is drawn into the joint by capillary action. Brazing results in a mechanical and metallurgical bond between work pieces.

## 2.2.3 Soldering

*13* Soldering is a method of joining metals below their melting points using a filler metal. Soldering, like brazing, results in an adhesive joint and occurs at lower temperatures than brazing, specifically below 450℃ (840 F).

## 2.3 Heat treatment

*14* Metals can be heat treated to alter the properties of strength, ductility, toughness, hardness or resistance to corrosion. Common heat treatment processes include annealing, precipitation strengthening, quenching, and tempering. The annealing process softens the metal by allowing recovery of cold work and grain growth. Quenching can be used to harden alloy steels, or in precipitation hardenable alloys, to trap dissolved solute atoms in solution. Tempering will cause the dissolved alloying elements to precipitate, or in the case of quenched steels, improve impact strength and ductile properties.

## 2.4 Surface treatment

### 2.4.1 Plating

*15* Electroplating is a common surface-treatment technique. It involves bonding a thin layer of another metal such as gold, silver, chromium or zinc to the surface of the product. It is used to reduce corrosion as well as to improve the product's aesthetic appearance.

### 2.4.2 Thermal spraying

*16* Thermal spraying techniques are another popular finishing option, and often have better high temperature properties than electroplated coatings.

### 2.4.3 Case hardening

*17* Case hardening is a process in which an alloying element, most commonly carbon or nitrogen, diffuses into the surface of a monolithic metal. The resulting interstitial solid solution is harder than the base material, which improves wear resistance without sacrificing toughness.

## 3. Metallurgical techniques

*18* Metallography allows metallurgists to study the microstructure of metals.

*19* Metallurgists study the microscopic and macroscopic properties using metallography, a technique invented by Henry Clifton Sorby. In metallography, an alloy of interest is ground flat and polished to a mirror finish. The sample can then be etched to reveal the microstructure and macrostructure of the metal. A metallurgist can then examine the sample with an optical or electron microscope and learn a great deal about the sample's composition, mechanical properties, and processing history.

Fig. 1.2 - 2 **The microstructure of metals**

20　Crystallography, often using diffraction of x-rays or electrons, is another valuable tool available to the modern metallurgists. Crystallography allows the identification of unknown materials and reveals the crystal structure of the sample. Quantitative crystallography can be used to calculate the amount of phases present as well as the degree of strain to which a sample has been subjected.

21　The physical properties of metals can be quantified by mechanical testing. Typical tests include tensile strength, compressive strength, hardness, impact toughness, fatigue and creep life.

## ❖ New Words and Phrases

| | |
|---|---|
| extractive [iks'træktiv] adj. | 抽取的 |
| extractive metallurgy | 提取冶金,冶炼 |
| joining ['dʒɔiniŋ] n. | 连接 |
| welding ['weldiŋ] n. | 焊接 |
| brazing [breiziŋ] n. | 硬钎焊 |
| soldering ['sɔldəriŋ] n. | 软钎焊 |
| heat treatment | 热处理 |
| surface treatment | 表面处理 |
| plating ['pleitiŋ] n. | 电镀 |
| refining [ri'fainiŋ] n. | 精炼 |
| sulfide ['sʌlfaid] n. | 硫化物 |
| electrolytically [i‚lektrəu'litikəli] adv. | 以电解 |
| metallurgist [me'tælədʒist] n. | 冶金家,冶金学家 |
| concentrate ['kɔnsentreit] v. | 浓缩,富集 |
| tailings ['teiliŋz] n. | 残渣,尾矿 |
| toughness ['tʌfnis] n. | 韧性 |
| hardness ['hɑːdnis] n. | 硬度 |

## Lesson 2  Metallurgy

| | |
|---|---|
| cryogenic [ˌkraiəu'dʒenik] a. | 低温学的；低温实验法的；制冷的，冷冻的 |
| cyclic ['saiklik] a. | 循环的 |
| cyclic loading | 周期载荷 |
| creep [kri:p] n. /vi | 蠕变 |
| rolling ['rəuliŋ] n. | 轧制 |
| extrusion [eks'tru:ʒən] n. | 挤出 |
| sintering ['sintəriŋ] n. | 烧结 |
| lathe [leið] n. | 车床 |
| milling ['miliŋ] n. | 磨 |
| milling machine n. | 铣床 |
| drill [dril] n. | 钻床 |
| guillotine ['giləti:n] n. | 轧刀，裁切机，剪床 |
| annealing ['a:ni:liŋ] n. | 退火 |
| precipitation [priˌsipi'teiʃən] n. | 坠落，沉淀 |
| precipitation strengthening | 析出强化 |
| quenching ['kwentʃiŋ] n. | 淬火 |
| tempering ['tempəriŋ] n. | 回火 |
| nitrogen ['naitrədʒən] n. | 氮 |
| interstitial [ˌintə(:)'stiʃəl] adj. | 组织间隙的，间质的 |
| interstitial solid solution | 间隙固溶体 |
| metallography [ˌmetə'lɔgrəfi] n. | 金属组织学，金相学 |
| microstructure [ˌmaikrəu'strʌktʃə] n. | 微观结构 |
| macrostructure [ˌmækrəu'strʌktʃə] n. | 宏观结构 |

## ❖ Notes

1. Metallurgy is a domain of materials science that studies the physical and chemical behavior of metallic elements, their intermetallic compounds, and their mixtures, which are called alloys.

   冶金学属于材料科学领域，是研究金属元素、金属间化合物及其混合物（即合金）的物理和化学特性的科学。

2. After mining, large pieces of the ore feed are broken through crushing and/or grinding in order to obtain particles small enough where each particle is either mostly valuable or mostly waste.

   开采后，为了获得足够小颗粒，大块的矿石被破碎和/或碾碎，基本上使得每个颗粒或者有用或者成为废料。

**语法补充:either…or…的用法**

**一、本语法在注释二中的应用**

either...or... 指二者中只能选择一个。本句中"each particle is either mostly valuable or mostly waste"是指每种颗粒的情况只能是"有价值"和"废料"中的一种。

**二、对本语法的详细阐述**

主要用于表示选择,其意为"要么……要么……""或者……或者……",用于连接两个性质相同的词或短语。

e.g. You can have either this one or that one.

你拿这个或那个都可以。

使用 either...or... 的注意点

a) either...or... 连接两个成分作主语时,谓语动词通常与其靠近的主语保持一致。

e.g. Either you or I am to go.

你或我必须有人去。

b) either...or... 除可连接两个词或短语外,有时也可连接两个句子。

e.g. Either you must improve your work or I shall dismiss you.

要么你改进工作,要么我就辞退你。

c) either...or... 的否定式可以是 not either...or...,也可以是 neither...nor...。

e.g. He didn't either write or phone. = He neither wrote nor phoned.

他既没写信又没打电话。

3. Metals exposed to cold or cryogenic conditions may endure a ductile to brittle transition and lose their toughness, becoming more brittle and prone to cracking.

   金属处在寒冷或低温环境下,可能会发生韧脆转变,韧性降低,脆性增加,很容易开裂。

4. Thermal spraying techniques are another popular finishing option, and often have better high temperature properties than electroplated coatings.

   热喷涂技术是另一种常见的最终表面处理技术,其涂层的高温性能往往比电镀层更好。

5. A metallurgist can then examine the sample with an optical or electron microscope and learn a great deal about the sample's composition, mechanical properties, and processing history.

   接下去,冶金学家通过光学或电子显微镜观察样品,了解样品的组成、力学性能和加工过程等情况。

# Lesson 2  Metallurgy

## *Check your understanding*

Ⅰ. Give brief answers to the following questions.
1. What is metallurgy?
2. How are metals shaped?
3. What is welding?

Ⅱ. Match the items listed in the following two columns.

| | |
|---|---|
| toughness | 电镀 |
| extractive metallurgy | 韧性 |
| plating | 焊接 |
| rolling | 冶炼 |
| drill | 微观结构 |
| welding | 钻床 |
| guillotine | 轧制 |
| microstructure | 车床 |
| lathe | 轧刀 |

## 冶 金 学

  冶金学属于材料科学领域,是研究金属元素、金属间化合物及其混合物(即合金)的物理和化学特性的科学,也是金属工艺学,生产中的科学,常用于金属加工工艺方面。

### 1. 冶炼

  冶炼是指从矿石里提取有价值的金属和精炼初步提取的金属达到较高的纯度。为了将金属氧化物或硫化物转换为纯金属,必须用物理、化学或电解的方法提炼矿石。

  右图是公元 1313 年,中国元代由王祯编著的《农书》所绘制的水车驱动的鼓风炉的图谱。

  冶金学家对这三大方面感兴趣:给矿,精矿(有价值的金属氧化物/硫化物)和尾矿(废弃物)。开采后,为了获得足够小的颗粒,大块的

图 1.2-1  水车驱动的鼓风炉

矿石被破碎或碾碎,基本上使得每个颗粒或者有用或者成为废料。富集有价值的颗粒,使得所需的金属能从废品中分离。

如果矿体和矿源容易被浸出,挖掘可能就没有必要。浸溶矿体里的矿物质并使之在溶液中富集。收集溶液并处理以提取有价值的金属。

矿体常常包含一种以上的有用金属。先前加工过程中的尾矿可能会被用作另一提炼过程的原料,从原矿石中提取次要产品。此外,富集可能含有一种以上的有用金属。随后将有价值金属分离开来。

**2. 金属的生产过程**

冶金学应用在为消费者或机械产品所使用的金属元件的生产过程中,涉及到合金的生产,成型,热处理和产品的表面处理。冶金学家的任务是实现材料属性(如成本,重量,强度,韧性,硬度,耐腐蚀和抗疲劳性等)和极限温度下性能的平衡。为了达到这一目标,必须认真加以考虑生产环境。在盐水环境中,黑色金属和某些铝合金腐蚀迅速。金属处在寒冷或低温环境下,可能会发生韧脆转变,韧性降低,脆性增加,很容易开裂。金属在连续循环载荷作用下会引起疲劳。金属在恒定应力下高温作业时可发生蠕变。

2.1 金属的加工工艺

金属成形工艺有铸造,锻造,流动旋压,轧制,挤压,烧结,金属加工,机械加工和制作。铸造成形时,熔融金属倒入铸模内成形。锻造时,炽热钢坯锤打成形。轧制是指钢坯连续通过窄轧辊变成薄板。挤压时,具有延展性的热金属是在压力下通过模具,在其冷却前成形。烧结是指,金属粉末在高温下通过模具压缩成形。机械加工是指用车床、铣床和钻床切削金属成形。制造成形是将金属片用截切机或气体切割机切断后弯曲成形。

"冷加工",如轧制、制作或其他工艺成形过程中,只要被加工的对象是冷的,就可通过加工硬化来提高产品的强度。加工硬化生成了金属的微观缺陷,它抵制进一步发生形变。

工业界和学术界应用的铸造各种各样,它们包括砂型铸造,熔模铸造(也称为"失蜡法"),压铸和连铸。

2.2 连接

2.2.1 焊接

焊接是通过加热使分离的物体连接成整体的技术。组成类似的填充料(焊条)也可融入连接处。

2.2.2 硬钎焊

硬钎焊是一种在低于基体熔点的温度下胶接金属的技术。使用熔点低于基体金属的钎料,利用毛细引力使液态钎料填充接头处。钎焊实现连接件间的机械和冶金连接。

2.2.3 软钎焊

软钎焊是一种加入焊料,在低于被连接件熔点的温度下连接金属的技术。软钎焊,和硬钎焊一样,形成胶粘联合,且发生在较硬钎焊低的温度下,尤其是低于450℃(840F)。

2.3 热处理

热处理可以改变金属的强度,塑性,韧性,硬度和耐腐蚀性等性能。普通的热处理工艺包括退火,沉淀强化,淬火,回火。退火是通过允许冷变形回复和晶粒结晶的方式来软化金属的工艺。淬火能使固溶体析出溶质原子,用于强化合金,或沉淀硬化合金。回火将导致溶解的合金元素析出,对淬火钢而言可提高抗冲击强度和韧性。

2.4 表面处理

2.4.1 电镀

电镀是一种常见的表面处理技术。它涉及在产品的表面粘接另一种金属薄层,如黄金、白银、铬或锌,用来降低腐蚀以及提高产品的美观性。

2.4.2 热喷涂

热喷涂技术是另一种常见的最终表面处理技术,其涂层的高温性能往往比电镀层更好。

2.4.3 表面硬化

表面硬化的工艺中,一种合金元素,最常见的碳或氮,扩散到整块金属的表面。由此产生的间隙固溶体的硬度比基体材料高,提高了耐磨性,而不牺牲韧性。

## 3. 冶金技术

冶金学家利用金相学研究金属的微观结构。

冶金学家使用金相研究材料的微观和宏观性能,金相由亨利·克利夫顿·索尔贝发明。在金相研究中,研磨合金的平面后抛光成镜面。然后侵蚀样品,可见金属的微观结构和宏观结构。接下去,冶金学家通过光学或电子显微镜观察样品,了解样品的组成、力学性能和加工过程等情况。

图 1.2-2 金属的微观结构

对现代冶金学家而言,晶体学是另一种宝贵的工具,晶体学利用 X 射线衍射或电子衍射作为研究手段。晶体学可以鉴定未知材料和揭示样品的晶体结构。定量晶体学可用于计算存在相的量,以及样品的应变程度。

金属的物理性能可以通过力学性能试验来量化研究。典型的试验包括抗拉强度,抗压强度,硬度,冲击韧性,疲劳和蠕变寿命的测试。

## Lesson 3  Metals

### 1. Properties of metals

1  The five most used metals are:
2    1) Iron
3    2) Aluminum
4    3) Copper
5    4) Zinc
6    5) Magnesium
7  The general physical properties of metals are:
8  They are strong and hard.
9  They are solids at room temperature (except for Mercury, which is the only metal to be liquid at room temperature)

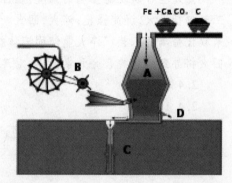

Fig. 1.3-1  Process of smelting of iron cannons

10  They have a shiny luster when polished.
11  They make good heat conductors and electrical conductors.
12  They are dense
13  They produce a sonorous sound when struck.
14  They have high melting points
15  They are malleable
16  The properties of metals make them suitable for different uses in daily life.
17  Copper is a good conductor of electricity and is ductile. Therefore Copper is used for electrical cables.
18  Gold and Silver are very malleable, ductile and very nonreactive. Gold and silver are used to make intricate jewelry which does not tarnish. Gold can also be used for electrical connections.
19  Iron and Steel are both hard and strong. Therefore they are used to construct bridges and buildings. A disadvantage of using iron is that it tends to rust, whereas most steels rust, but they can be formulated to be rust free.
20  Aluminum is a good conductor of heat and is malleable. It is used to make saucepans and tin foil, and also aeroplane bodies as it is very light.

*21* Pure elemental metals are often too soft to be of practical use which is why much of metallurgy focuses on formulating useful alloys.

## 2. Important common alloy systems

*22* Common engineering metals include aluminium, chromium, copper, iron, magnesium, nickel, titanium and zinc. These are most often used as alloys. Much effort has been placed on understanding the iron-carbon alloy system, which includes steels and cast irons. Plain carbon steels are used in low cost, high strength applications where weight and corrosion are not a problem. Cast irons, including ductile iron are also part of the iron-carbon system.

*23* Stainless steel or galvanized steel are used where resistance to corrosion is important. Aluminium alloys and magnesium alloys are used for applications where strength and lightness are required.

*24* Cupro-nickel alloys such as Monel are used in highly corrosive environments and for non-magnetic applications. Nickel-based superalloys like Inconel are used in high temperature applications such as turbochargers, pressure vessels, and heat exchangers. For extremely high temperatures, single crystal alloys are used to minimize creep.

## 3. Chemical properties of metals

*25* The black surface is lead oxide. The white surface is the lead viewed when the lead oxide is scratched.

*26* Substances on the Earth's surface will come in contact with air, water or acids. A major concern for the use of metals is their corrosion. The shiny surface of many metals becomes dull in time. This is due to a slow chemical reaction between the surface of the metal and oxygen in the air; that is typically a surface coating of the metal oxide. The general word equation is:

Fig. 1.3 – 2  **A block of metal after corrosion**

*27* metal + oxygen ⟶ metal oxide

*28* For example: The dull appearance of the metal lead is due to a coating of lead oxide.

*29* lead + oxygen ⟶ lead oxide

*30* If the surface is scratched, then the shiny lead metal can be seen underneath.

*31* Heating can speed up the reaction with oxygen. If a piece of copper is heated it quickly becomes coated in black copper oxide. The word equation is:

*32* copper + oxygen ⟶ copper oxide

## ❖ New Words and Phrases

| | |
|---|---|
| copper [ˈkɔpə] n. | 铜 |
| zinc [ziŋk] n. | 锌 |
| magnesium [mægˈniːzjəm] n. | 镁 |
| luster [ˈlʌstə] n. | 光泽 |
| polish [ˈpɔliʃ] v. | 抛光 |
| conductor [kənˈdʌktə] n. | 导体 |
| heat conductor | 导热体 |
| electrical conductor | 导电体 |
| malleable [ˈmæliəbl] a. | 可锻造的，有延展性的，韧性的 |
| ductile [ˈdʌktail] a. | 可延展的，有韧性的 |
| intricate [ˈintrikit] a. | 错综复杂的 |
| tarnish [ˈtɑːniʃ] v. | 金属失去光泽；n. 无光泽 |
| rust [rʌst] n. | 锈 v.(使)生锈 |
| alloy [ˈælɔi] n. | 合金 |
| alloy system | 合金系 |
| corrosion [kəˈrəuʒən] n. | 腐蚀,受腐蚀的部位 |
| stainless [ˈsteinlis] adj. | 不锈的 |
| stainless steel | 不锈钢 |
| galvanized [ˈgælvənaizd] adj. | 镀锌的 |
| galvanized steel | 镀锌钢 |
| cupro-nickel alloys | 铜镍合金 |
| magnetic [mægˈnetik] adj. | 有磁性的 |
| non-magnetic a. | 无磁性的 |
| superalloy [ˌsjuːpəˈælɔi] n. | 超耐热合金,超级合金,高温高强度合金 |
| inconel [ˈinkəunəl] n. | 铬镍铁合金,因康镍合金 |
| turbocharger [ˈtəːbəuˌtʃɑːdʒə] n. | 涡轮增压器 |
| vessel [ˈvesl] n. | 容器 |
| pressure vessel | 压力容器 |
| strength [streŋθ] n. | 强度 |

## ❖ Notes

1. Pure elemental metals are often too soft to be of practical use which is why much of

# Lesson 3    Metals

metallurgy focuses on formulating useful alloys.

通常,纯金属因太软而不实用,因此冶金重点多放在制备有用的合金上。

**语法补充:引导表语从句的引导词**

一、本语法在注释一中的应用

　　所谓表语从句是指接在 be 动词或系动词后面的从句。注释一中"why much of metallurgy focuses on formulating useful alloys"是一个由关系副词 why 引导的表语从句,说明了冶金重点不放在纯金属上的原因。

二、对本语法的详细阐述

(一) why 引导表语从句是针对前面已经说明过的原因进行总结,"That is why..."是表达这种意思的常用句型,意为"这就是……的原因/因此……"。

e.g. That is why you see this old woman before you know, Jeanne.

　　　珍妮,这就是现在这个老太婆出现在你面前的原因。

(二) that, why 与 because 引导表语从句的区别

　　that 没有词义,只起连接作用;why 强调结果; because 强调原因。

e.g. The reason was that you don't trust her.

　　　原因是你不信任她。

　　　The fact is that they are angry with each other.

　　　事实是他们生彼此的气。

　　　He was ill. That's why he was sent to the hospital.

　　　他病了,所以被送到医院来。

　　　He was sent to the hospital. That's because he was ill.

　　　他被送到医院,是因为他生病了。

## Check your understanding

Ⅰ. Give brief answers to the following questions.

　　1. What are the five most used metals?

　　2. Which metal is used for electrical cables?

　　3. What features does aluminum have?

Ⅱ. Match the items listed in the following two columns.

　　alloy　　　　　　　　　　　锌

　　zinc　　　　　　　　　　　 铬镍铁合金

　　heat conductor　　　　　　不锈钢

　　galvanized steel　　　　　 压力容器

| inconel | 导热体 |
| magnetic | 镀锌钢 |
| stainless steel | 合金 |
| pressure vessel | 有磁性的 |

# 金 属

## 1. 金属的性能

五种最常用的金属是：
1. 铁
2. 铝
3. 铜
4. 锌
5. 镁

金属普遍的物理性能是：

具有强度和硬度；

在室温下是固体（汞除外，汞是常温下唯一的液态金属）；

经抛光后具有金属光泽；

是热和电的良导体；

致密；

敲击时发出铿锵声；

熔点高；

具有可塑性。

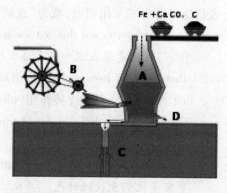

图1.3-1 加农炮高速钢的熔炼过程

金属的特性使它们在日常生活中有着不同的用途。

铜是电的良导体且具有韧性。因此，铜适用于电缆。

金和银韧性、塑性非常好，不易发生反应。黄金和白银可用来制造复杂形状、保持金属光泽的首饰。黄金还可以用于电气连接。

钢铁具有硬度和强度。因此，它们被用来建造桥梁和建筑物。铁在使用时往往存在生锈的缺点。大部分钢也会生锈，但可以通过改变配方来避免生锈。

铝具有良好的导热性和可塑性，常用来制造锅和锡箔；因为它很轻，因此也用来制造飞机的部件。

通常，纯金属因太软而不实用，因此冶金重点多放在制备有用的合金上。

## 2. 常用的重要的合金系

常用工程金属包括铝,铬,铜,铁,镁,镍,钛和锌。它们最常用作合金。人们很重视研究铁碳合金系,包括钢和铸铁。普通碳钢一般用在成本低,强度高,而不考虑重量和腐蚀的场合。铸铁,包括球墨铸铁也是铁碳合金体系的一部分。

不锈钢或镀锌钢板用于需要耐腐蚀的场合。铝合金和镁合金适用于对强度和质轻有要求的场合。

铜镍合金,如蒙奈尔铜镍合金,用于高腐蚀性的或无磁性环境。镍基高温合金如因康镍合金,在高温下使用,如涡轮增压器,压力容器和热交换器。单晶合金在极高的温度下使用,以尽量减少蠕变。

图1.3-2 腐蚀后的金属
黑色表面是氧化铅。
白色的表面是刮去氧化铅看到的金属铅

## 3. 金属的化学性质

物质在地球表面会接触到空气,水或酸。金属在使用时会腐蚀,这是人们关心的一个主要问题。许多有光泽的金属表面随时间而变得晦暗。这是因为金属的表面和空气中的氧气发生了缓慢的化学反应,这是典型的表面金属氧化物的覆盖层。一般方程是:

金属 + 氧气 → 金属氧化物

例如:金属铅由于有氧化铅覆盖层而呈现晦暗的外观。

铅 + 氧气 → 氧化铅 如果刮去表面氧化铅,可以看到下面带金属光泽的铅。加热可以加快金属与氧气的反应。铜在加热时,它表面将被氧化为黑色氧化铜。一般方程是:铜 + 氧气 → 氧化铜

## Lesson 4  Materials in Industry

*1*  Radical materials advances can drive the creation of new products or even new industries, but stable industries also employ materials scientists to make incremental improvements and troubleshoot issues with currently used materials. Industrial applications of materials science include materials design, cost-benefit tradeoffs in industrial production of materials, processing techniques (casting, forging, welding, ion implantation, crystal growth, thin-film deposition, sintering, glassblowing, etc.) and analytical techniques [characterization techniques such as electron microscopy, x-ray diffraction, thermal analysis, nuclear microscopy (HEFIB), Rutherford backscattering, neutron diffraction, small-angle X-ray scattering (SAXS], etc..

*2*  Besides material characteristics, the material scientists/engineers also deal with the extraction of materials and their conversion into useful forms. Thus ingot casting, foundry techniques, blast furnace extraction, and electrolytic extraction are all parts of the required knowledge of a metallurgist/engineer. Often the presence, absence or variation of minute quantities of secondary elements and compounds in a bulk material will have a great impact on the final properties of the materials produced, for instance, steels are classified based on weight percentages of the carbon and other alloying elements they contain. Thus, the extraction and purification techniques employed in the extraction of iron in the blast furnace will have an impact on the quality of steel that may be produced.

*3*  The overlap between physics and materials science has led to the offshoot field of materials physics, which is concerned with the physical properties of materials. The approach is generally more macroscopic and applied than in condensed matter physics.

*4*  The study of metal alloys is a significant part of materials science. Of all the metallic alloys in use today, the alloys of iron (steel, stainless steel, cast iron, tool steel, alloy steels) make up the largest proportion both by quantity and commercial value. Iron alloyed with various proportions of carbon gives low, mid and high carbon steels. For the steels, the hardness and tensile strength of the steel is directly related to the amount of carbon present, with increasing carbon levels also leading to lower ductility and toughness. The addition of silicon and graphitization will produce cast irons (although some cast irons are made precisely with no graphitization). The addition of chromium, nickel and molybdenum to carbon steels (more than 10%) gives us stainless steels.

# Lesson 4    Materials in Industry

5    Other significant metallic alloys are those of aluminum, titanium, copper and magnesium alloys. Copper alloys have been known for a long time (since the Bronze Age), while the alloys of the other three metals have been relatively recently developed. Due to the chemical reactivity of these metals, the electrolytic extraction processes required were only developed relatively recently. The alloys of aluminum, titanium and magnesium are also known and valued for their high strength-to-weight ratios and, and in the case of magnesium, their ability to provide electromagnetic shielding. These materials are ideal for situations where high strength-to-weight ratios are more important than bulk cost, such as in the aerospace industry and certain automotive engineering applications.

6    Other than metals, polymers and ceramics are also important parts of materials science. Polymers are the raw materials (the resins) used to make what we commonly call plastics. Plastics are really the final products, created after one or more polymers or additives have been added to a resin during processing, which is then shaped into a final form. Polymers which have been around, and which are in current widespread use, include polyethylene, polypropylene, PVC, polystyrene, nylons, polyesters, acrylics, polyurethanes, and polycarbonates. Plastics are generally classified as "commodity", "specialty" and "engineering" plastics.

7    PVC (polyvinyl-chloride) is widely used, inexpensive, and annual production quantities are large. It lends itself to an incredible array of applications, from artificial leather to electrical insulation and cabling, packaging and containers. Its fabrication and processing are simple and well-established. The versatility of PVC is due to the wide range of plasticisers and other additives that it accepts. The term "additives" in polymer science refers to the chemicals and compounds added to the polymer base to modify its material properties.

8    Polycarbonate would be normally considered an engineering plastic (other examples include PEEK, ABS). Engineering plastics are valued for their superior strengths and other special material properties. They are usually not used for disposable applications, unlike commodity plastics.

9    Specialty plastics are materials with unique characteristics, such as ultra-high strength, electrical conductivity, electro-fluorescence, high thermal stability, etc.

10    It should be noted here that the dividing line between the various types of plastics is not based on material but rather on their properties and applications. For instance, polyethylene (PE) is a cheap, low friction polymer commonly used to make disposable shopping bags and trash bags, and is considered a commodity plastic, whereas Medium-Density Polyethylene MDPE is used for underground gas and water pipes, and another

variety called Ultra-high Molecular Weight Polyethylene UHMWPE is an engineering plastic which is used extensively as the glide rails for industrial equipment and the low-friction socket in implanted hip joints.

*11* Another application of material science in industry is the making of composite materials. Composite materials are structured materials composed of two or more macroscopic phases. An example would be steel-reinforced concrete; another can be seen in the "plastic" casings of television sets, cell-phones and so on. These plastic casings are usually a composite material made up of a thermoplastic matrix such as acrylonitrile-butadiene-styrene (ABS) in which calcium carbonate chalk, talc, glass fibres or carbon fibres have been added for added strength, bulk, or electro-static dispersion. These additions may be referred to as reinforcing fibres, or dispersants, depending on their purposes.

❖ **New Words and Phrases**

| | |
|---|---|
| tradeoff ['treid,ɔːf] n. | 折中,权衡 |
| diffraction [di'frækʃən] n. | 衍射 |
| ion ['aiən] n. | 离子 |
| implantation [,implɑːn'teiʃən] n. | 灌输 |
| ion implantation | 离子注入 |
| electrolytic [i,lektrəu'litik] adj. | 电解的 |
| extraction [iks'trækʃən] n. | 抽出,取出 |
| electrolytic extraction | 电解提取[分离,抽提,萃取] |
| titanium [tai'teinjəm] n. | 钛 |
| graphitization [,græfitai'zeiʃən] n. | 石墨化 |
| chromium ['krəumjəm] n. | 铬 |
| nickel ['nikl] n. | 镍 |
| molybdenum [mə'libdinəm] n. | 钼 |
| magnesium [mæg'niːzjəm] n. | 镁 |
| polyethylene [,pɔli'eθiliːn] n. | 聚乙烯 |
| polypropylene [,pɔli'prəupiliːn] n. | 聚丙烯 |
| polystyrene [,pɔli'staiəriːn] n. | 聚苯乙烯 |
| nylon ['nailən] n. | 尼龙 |
| polyester ['pɔliestə] n. | 聚酯纤维,涤纶 |
| acrylics [ə'kriliks] n. | 丙烯酸树脂 |
| acrylonitrile [,ækrələu'naitril] n. | 丙烯腈(氰乙烯) |

# Lesson 4　Materials in Industry

styrene [ˈstairiːn] n.　　　　　　　　苯乙烯
polyurethane [ˌpɔliˈjuəriθein] n.　　聚氨酯
polycarbonate [ˌpɔliˈkɑːbənit] n.　　聚碳酸酯
dispersant [disˈpəːsənt] n.　　　　　分散剂

## ❖ Notes

1. Often the presence, absence or variation of minute quantities of secondary elements and compounds in a bulk material will have a great impact on the final properties of the materials produced, ...
   往往,微量元素及化合物存在与否或其量的变化都将对材料的最终性能产生巨大影响,……

2. For the steels, the hardness and tensile strength of the steel is directly related to the amount of carbon present, with increasing carbon levels also leading to lower ductility and toughness.
   对于钢而言,含碳量直接影响钢的硬度和拉伸强度,随含碳量的增加,塑性和韧性降低。

3. It should be noted here that the dividing line between the various types of plastics is not based on material but rather on their properties and applications.
   这里应该指出,不同类型的塑料的分界线并非基于材料,而是基于其性能及应用。

**语法补充:It 做形式主语的用法**

一、本语法在注释三中的应用
　　在主语太长时,可以用 it 做形式主语,从来避免句子结构"头重脚轻"。注释三中 it 是形式主语,真正的主语是"that the dividing line between the various types of plastics is not based on material but rather on their properties and applications"。

二、对本语法的详细阐述
1. It ＋ be ＋ 形容词＋主语从句
   e.g. It is uncertain whether he can come to Jenny's birthday party or not.
   　　不知道他能不能来参加珍妮的生日聚会。
2. It ＋ be ＋ 名词词组 ＋ 主语从句
   e.g. It's a pity that you missed the exciting football match.
   　　你错过这场激动人心的足球比赛是个遗憾。
3. It ＋ be ＋ 过去分词 ＋ 主语从句
   e.g. It is reported that 16 people were killed in the earthquake.
   　　据报道,16 个人在这场地震中丧生。

4. It + 不及物动词(seem, appear, happen 等) + 主语从句

e. g. It seemed that he didn't tell the truth.

  似乎他没有说实话。

5. It + be + 形容词 +( for sb. ) + 动词不定式

有时候为了强调不定式动作的执行者,常在不定式前加 for sb。

e. g. It's necessary for the young to master two foreign languages.

  年轻人掌握两门外语是必要的。

e. g. It is unwise to give the children whatever they want.

  孩子想要什么就给什么是不明智的。

6. It + be + 形容词 + of sb. + 动词不定式,这类形容词常是表示心理品质,性格特征的形容词。如:kind, nice, stupid, clever, foolish, polite, impolite, silly, selfish, considerate 等。

e. g. It's very kind of you to help me with the work.

  你帮我工作,你真好。

7. It + be +名词词组 + 动词不定式

e. g. It is not a good habit to stay up too late.

  熬夜不是好习惯。

8. It + be + 名词或形容词 + 动名词

e. g. It's a waste of time talking to her any more.

  你再和她谈就是浪费时间。

9. It + take (sb. ) +时间(金钱) + 动词不定式

e. g. It took the workers almost three years to finish building the dam.

  工人们花费了差不多三年才建好这个水坝。

## Check your understanding

Ⅰ. Give brief answers to the following questions.

 1. What are the industrial applications of materials science?

 2. For the steels, what directly affects the hardness and tensile strength of the steel?

 3. What are the composite materials?

Ⅱ. Match the items listed in the following two columns.

  ion         尼龙

  polyethylene     电解提取

  electrolytic extraction  钛

  nickel        离子

| nylon | 聚乙烯 |
| titanium | 分散剂 |
| dispersant | 聚酯纤维，涤纶 |
| polyester | 镍 |

## 材料在工业中的应用

材料的重大发展可以推动新产品、甚至新产业的诞生，但稳定的行业也需要材料科学家对目前使用的材料进行逐步改进并排解故障。材料科学的工业应用包括材料设计，工业生产中成本效益权衡，加工工艺（铸造、锻造、焊接、离子注入、晶体生长、薄膜沉积、烧结、玻璃吹制等）和分析技术[表征技术，如电子显微镜、X射线衍射、热分析、核显微镜（HEFIB）、卢瑟福背散射、中子衍射、小角度X射线散射（SAXS）]等等。

除了研究材料特性外，材料科学家/工程师还研究材料的提炼及其向有用形式的转化。因此铝锭铸造，铸造技术，高炉提炼和电解提取等都是冶金学家/工程师必要的知识。往往，微量元素及化合物存在与否或其量的变化都将对材料的最终性能产生巨大影响，例如，钢是以其含碳和其他合金元素的质量百分数来分类的。因此，高炉炼铁中采用的冶炼和纯化技术将影响生产的钢铁质量。

材料科学与物理学交叉衍生出研究材料物理性质的分支学科——材料物理。比之凝聚态物理，材料物理通常更侧重于宏观和应用。

对金属合金的研究是材料科学的一个重要组成部分。在现在所用的合金中，铁合金（钢、不锈钢、铸铁、工具钢、合金钢）在数量和经济价值方面具有最大的份额。含碳量不同的铁合金可以分为低、中、高碳钢。对于钢而言，含碳量直接影响钢的硬度和拉伸强度，随含碳量的增加，塑性和韧性降低。铸铁生产需添加硅和石墨化处理（虽然有些铸铁恰恰是没有石墨化处理）。添加（超过10%）铬、镍和钼到碳钢中将形成不锈钢。

其他重要的金属合金是铝、钛、铜和镁合金。（自铜器时代）铜合金已被知道了很长一段时间，相对而言，其他三个金属合金是最近才研发的。由于这些金属的化学反应，所需要的电解提取工艺研究相对较晚。铝、钛，镁合金因为具有高的比强度而知名并得到重视，镁合金还具有电磁屏蔽能力。这些材料在应用于高比强度重要于高成本的场合下是理想的，如在航空航天工业和某些汽车工程应用领域。

除了金属，聚合物和陶瓷材料也是材料科学的重要组成部分。聚合物是用于制造我们熟知的塑料的原料（树脂）。塑料实际上是最终产品，在其加工过程中一种或多种聚合物或添加剂加入树脂，然后最终成形。目前，广泛使用的聚合物，包括聚乙烯、聚丙烯、聚氯乙烯、聚苯乙烯、尼龙、聚酯、丙烯酸树脂、聚氨酯和聚碳酸酯。塑料一般分为"普通"、"特种"和"工程"塑料。

广泛使用的、廉价的 PVC(聚氯乙烯)年产量很大。其应用范围广泛得令人难以置信,从人造皮革、电绝缘材料及电缆到包装材料和容器。它的制造和加工简单且行之有效。PVC 的多用性归功于它可接受各种各样的增塑剂和添加剂。在高分子科学中的术语"添加剂",是指为改变材料性能而添加到聚合物基体中的化学品和化合物。

聚碳酸酯通常被认为是工程塑料(其他例子包括聚醚醚酮,ABS 树脂)。工程塑料因优越的强度和其他特性而得到重视。与普通塑料不同,工程塑料一般不用于生产一次性用品。

特种塑料具有独特的特点,如超高强度,导电性,电荧光性,较高的热稳定性等。

这里应该指出,不同类型的塑料的分界线并非基于材料,而是基于其性能及应用。例如,聚乙烯(PE)是一种廉价的,低摩擦聚合物,常用作一次性购物袋和垃圾袋,并且被视为一种普通塑料;中密度聚乙烯用于地下燃气及水的管道,而另一种被誉为超高分子量的工程塑料聚乙烯,广泛用作工业设备的滑动导轨和植入髋关节的低摩擦骨白。

材料科学的另一种应用是在复合材料的工业制造领域。复合材料是由两种或两种以上的宏观相组成的结构材料。例如,钢筋混凝土;再如电视机、手机等的"塑料"外壳。这些塑料外壳通常是热塑性基体的复合材料组成的,如丙烯腈,丁二烯,苯乙烯(ABS 树脂),其中添加了碳酸钙粉,滑石粉,玻璃纤维或碳纤维,以增加其强度、体积或静电分散性。这些增加剂可以称为强化纤维,或分散剂,这取决于它们的作用。

## 构词法之派生法

一、定义:派生法指通过在词根上加前缀或者后缀构成一个新词

二、具体构成方式

(一)前缀

1. 只改变词意,不改变词性

1) nano-:与"纳米"有关

e. g. nanoscience:[名词]纳米技术

2) semi-:一半

e. g. semiconductor:[名词]半导体

3) thermo-:与"热"有关

e. g. thermoplastics:[名词]热塑性塑料

4) micro-:微型的

e. g. microstructure:[名词]微观结构

5) inter-:相互之间的

e. g. interaction:[名词]互动

6) un-, non-, in-, dis-, ir-:表示否定

## Lesson 4  Materials in Industry

e.g. uneconomical：[形容词]不划算的
   non-ferrous metal：[名词]有色金属
   inexpensive：[形容词]便宜的
   disadvantageous：[形容词]不利的
   irregular：[形容词]不规则的
7）re-：再次
e.g. remove：[动词]消除
2. 既改变词意又改变词性（只有少数前缀能如此）
en-：可以将形容词和名词变成动词。
e.g. 1）large：[形容词] 大的
       enlarge：[动词] 扩大
e.g. 2）cage：[名词] 笼子
       encage：[动词] 关在笼中

# Unit 2

## Lesson 1  Casting

1   Casting is a manufacturing process by which a liquid material is (usually) poured into a mold, which contains a hollow cavity of the desired shape, and then allowed to solidify. The solid casting is then ejected or broken out to complete the process. Casting may be used to form hot liquid metals or various materials that cold set after mixing of components (such as epoxy, concrete, plaster and clay). Casting is most often used for making complex shapes that would be otherwise difficult or uneconomical to make by other methods.

Fig. 2.1 - 1   **Casting**

2   Casting is a 6000-year-old process. The oldest surviving casting is a copper frog from 3200 BC.

3   The casting process is subdivided into two distinct subgroups: expendable and non-expendable mold casting.

### Expendable mold casting

4   Expendable mold casting is a generic classification that includes sand, plastic, shell, plaster, and investment (lost-wax technique) moldings. This method of mold casting involves the use of temporary, non-reusable molds.

### Waste molding of plaster

5   A durable plaster intermediate is often used as a stage toward the production of a bronze sculpture or as a pointing guide for the creation of a carved stone. With the completion of a plaster, the work is more durable (if stored indoors)

Fig. 2.1 - 2   **Casting iron in a sand mold**

than a clay original which must be kept moist to avoid cracking. With the low cost plaster at hand, the expensive work of bronze casting or stone carving may be deferred until a patron is found, and as such work is considered to be a technical, rather than artistic process, it may even be deferred beyond the lifetime of the artist.

6   In waste molding a simple and thin plaster mold, reinforced by sisal or burlap, is cast over the original clay mixture. When cured, it is then removed from the damp clay, incidentally destroying the fine details in undercuts present in the clay, but which are now captured in the mold. The mold may then at any later time (but only once) be used to cast a plaster positive image, identical to the original clay. The surface of this "plaster" may be further refined and may be painted and waxed to resemble a finished bronze casting.

## Sand casting

7   Sand casting is one of the most popular and simplest types of casting that has been used for centuries. Sand casting allows for smaller batches to be made compared with permanent mold casting and at a very reasonable cost. Not only does this method allow manufacturers to create products at a low cost, but there are other benefits to sand casting, such as very small size operations. From castings that fit in the palm of your hand to train beds (one casting can create the entire bed for one rail car), it can all be done with sand casting. Sand casting also allows most metals to be cast depending on the type of sand used for the molds.

8   Sand casting requires a lead time of days for production at high output rates (1-20 pieces/hr-mold) and is unsurpassed for large-part production. Green (moist) sand has almost no part weight limit, whereas dry sand has a practical part mass limit of 2300-2700 kg. Minimum part weight ranges from 0.075-0.1 kg. The sand is bonded together using clays (such as green sand) or chemical binders, or polymerized oils (such as motor oil). Sand can be recycled many times in most operations and requires little additional input.

## Plaster casting (of metals)

9   Plaster casting is similar to sand molding except that plaster is substituted for sand. Plaster compound is actually composed of 70-80% gypsum and 20-30% strengthener and water. Generally, the form takes less than a week to prepare, after which a production rate of 1-10 units/hr-mold is achieved, with items as massive as 45 kg and as small as 30 g with very high surface resolution and fine tolerances. Parts that are typically made by plaster casting are lock components, gears, valves, fittings, tooling, and ornaments. Plaster casting is an inexpensive alternative to other molding processes due to the low cost of the plaster and the mold production. It may be disadvantageous, however, because the mold quality is dependent on several factors, "including consistency of the plaster molding composition, mold pouring procedures, and plaster curing techniques." If these factors are

not closely monitored, the mold can result in distorted dimensions, shrinking upon drying and poor mold surfaces.

10  Once used and cracked away, normal plaster cannot easily be recast. Plaster casting is normally used for non-ferrous metals such as aluminium-, zinc-, or copper-based alloys. It cannot be used to cast ferrous material because sulfur in gypsum slowly reacts with iron. The plaster itself cannot stand temperatures above 1200℃, which also limits the materials to be cast in plaster. Prior to mold preparation the pattern is sprayed with a thin film of parting compound to prevent the mold from sticking to the pattern. The unit is shaken, so plaster fills the small cavities around the pattern. The plaster sets, usually in about 15 minutes, and the pattern is removed. The plaster is dried at temperatures between 120° and 260℃. The mold is preheated and the molten metal poured in.

11  Plaster casting represents a step up in sophistication and requires skill. The automatic functions are easily handed over to robots, yet the higher-precision pattern designs required demand even higher levels of direct human assistance.

## Casting of plaster, concrete, or plastic resin

12  Plaster itself may be cast, as can other chemical setting materials such as concrete or plastic resin — either using single-use waste molds as noted above or multiple-use piece molds, or molds made of small ridged pieces or of flexible material such as latex rubber (which is in turn supported by an exterior mold). When casting plaster or concrete, the finished product is, unlike marble, relatively unattractive, lacking in transparency, and so it is usually painted, often in ways that give the appearance of metal or stone. Alternatively, the first layers cast may contain colored sand so as to give an appearance of stone. By casting concrete, rather than plaster, it is possible to create sculptures, fountains, or seating for outdoor use. A simulation of high-quality marble may be made using certain chemically-set plastic resins (for example epoxy or polyester) with powdered stone added for coloration, often with multiple colors worked in. The latter is a common means of making attractive washstands, washstand tops and shower stalls, with the skilled working of multiple colors resulting in simulated staining patterns as is often found in natural marble or travertine.

## Shell molding

13  Shell molding is also similar to sand molding except that a mixture of sand and 3-6% resin holds the grains together. Shell molding also uses sand with a much smaller grain than green-sand. Set-up and production of shell mold patterns takes weeks, after which an

output of 5-50 pieces/hr-mold is attainable. Aluminium and magnesium products average about 13.5 kg as a normal limit, but it is possible to cast items in the 45-90 kg range. Shell mold walling varies from 3-10 mm thick, depending on the forming time of the resin.

*14*   Shell molding is used for small parts that require high precision. Some examples include gear housings, cylinder heads and connecting rods. It is also used to make high-precision moulding cores. This process makes it possible that such complex parts can be cast with less labor.

*15*   There are a dozen different stages in shell mold processing that include:

(1) Initially preparing a metal-matched plate

(2) Mixing resin and sand

(3) Heating pattern, usually to between 505-550 K

(4) Inverting the pattern (the sand is at one end of a box and the pattern at the other, and the box is inverted for a time determined by the desired thickness of the mill)

(5) Curing shell and baking it

(6) Removing investment

(7) Inserting cores

(8) Repeating for other half

(9) Assembling mold

(10) Pouring mold

(11) Removing casting

(12) Cleaning and trimming.

*16*   The sand-resin mix can be recycled by burning off the resin at high temperatures.

## Investment casting

*17*   Investment casting (known as lost-wax casting in art) is a process that has been practiced for thousands of years, with the lost-wax process being one of the oldest known metal forming techniques. From 5000 years ago, when bees wax formed the pattern, to today's high technology waxes, refractory materials and specialist alloys, the castings ensure high-quality components are produced with the key benefits of accuracy, repeatability, versatility and integrity.

Fig. 2.1-3  Valve for Nuclear Power Station produced using investment casting

*18*   Investment casting derives its name from the fact that the pattern is invested, or surrounded, with a refractory material. The wax patterns require extreme care for they are not strong enough to withstand forces encountered during the mold making. One advantage of investment casting is that the wax can be reused.

*19*　The process is suitable for repeatable production of net shape components from a variety of different metals and high performance alloys. Although generally used for small castings, this process has been used to produce complete aircraft door frames, with steel castings of up to 300 kg and aluminium castings of up to 30 kg. Compared with other casting processes such as die casting or sand casting, it can be an expensive process, however, the components that can be produced using investment casting can incorporate intricate contours, and in most cases the components are cast near net shape, so requiring little or no rework once cast.

## ❖ New Words and Phrases

| | |
|---|---|
| casting ['kɑ:stiŋ] n. | 铸造 |
| eject [i'dʒekt] vt. | 顶出 |
| break out | 铸漏 |
| mold [məuld] n. | 铸型 |
| temporary ['tempərəri] adj. | 一次性的 |
| expendable mold casting | 消失模铸造 |
| durable ['djuərəbl] adj. | 持久的,耐用的 |
| bronze [brɔnz] n. | 铜 |
| sculpture ['skʌlptʃə] n. | 雕塑 |
| carve [kɑ:v] vt. & vi. | 雕刻 |
| clay [klei] n. | 泥土 |
| defer [di'fə:] vt. | 拖延,推迟 |
| patron ['peitrən] n. | 资助人 |
| sisal ['sisəl] n. | 剑麻,西沙尔麻 |
| burlap ['bə:læp] n. | 粗麻布 |
| cure [kjuə] vt. | 愈合,凝固 |
| refine [ri'fain] vt. | 使变得完善 |
| resemble [ri'zembl] vt. | 类似于 |
| sand casting | 砂型铸造 |
| batch [bætʃ] n. | 一批,一组 |
| permanent ['pə:mənənt] adj. | 永久性的 |
| palm [pɑ:m] n. | 掌状物 |
| rail car | 机动轨道车 |
| mass [mæs] n. | 大量,大批 |

## Lesson 1　Casting

| | |
|---|---|
| bond [bɔnd] vt. | 使黏合,使结合 |
| binder ['baində] n. | 黏合物 |
| polymerized oils | 聚合油 |
| motor oil | 机油 |
| plaster casting | 石膏模铸造 |
| gypsum ['dʒipsəm] n. | 石膏 |
| tolerance ['tɔlərəns] n. | 公差 |
| gear [giə] n. | 齿轮 |
| valve [vælv] n. | 阀门 |
| fitting ['fitiŋ] n. | 日用器具 |
| tooling ['tu:liŋ] n. | 模具 |
| ornament ['ɔ:nəmənt] n. | 装饰品 |
| monitor ['mɔnitə] vt. | 检测 |
| distort [dis'tɔ:t] vt. | 使变形 |
| shrink [ʃriŋk] vt. & vi. | 收缩 |
| ferrous material | 黑色金属 |
| sulfur ['sʌlfə] n. | 硫 |
| pattern ['pætən] n. | 型芯 |
| thin film | 薄膜 |
| parting compound | 脱模剂 |
| stick [stik] vt. & vi. | 粘贴,卡住 |
| cavity ['kæviti] n. | 腔,洞 |
| sophistication [sə͵fisti'keiʃən] n. | 复杂性 |
| latex rubber | 橡胶 |
| exterior [eks'tiəriə] adj. | 外部的 |
| marble ['mɑ:bl] n. | 大理石 |
| transparency [træns'pɛərənsi] n. | 透明性 |
| simulation [͵simju'leiʃən] n. | 模拟 |
| washstand ['wɔʃstænd] n. | 盥洗盆、盥洗台 |
| shower stall | 淋喷头 |
| stain [stein] vt. & vi. | 使染色 |
| shell molding | 壳模铸造 |
| investment casting | 熔模铸造 |
| refractory [ri'fræktəri] adj. | 耐熔的 |
| alloy ['ælɔi] n. | 合金 |

specialist alloy　　　　　　　　　　专用合金
component [kəmˈpəunənt] n.　　部件，原件
accuracy [ˈækjurəsi] n.　　　　　精确性
repeatability [ripiːtəˈbiliti] n.　　可重复性
versatility [ˌvɜːsəˈtiləti] n.　　　通用性
integrity [inˈtegriti] n.　　　　　完整性
incorporate [inˈkɔːpəreit] vt.　　把…合并
intricate [ˈintrikit] adj.　　　　　错综复杂的
contour [ˈkɔntuə] n.　　　　　　轮廓

## ❖ Notes

1. Casting is a manufacturing process by which a liquid material is (usually) poured into a mold, which contains a hollow cavity of the desired shape, and then allowed to solidify. (一般而言,)铸造是将液体浇进铸型里,铸型是空心的,然后凝固成形得到预定形状的一种制造工艺过程。

**语法补充:介词 + which 的定语从句用法**
一、本语法在注释一中的应用
　　"by which a liquid material is (usually) poured into a mold"是一个由 which 引导的定语从句,其中 which 做 by 这个介词的宾语,which 指代 casting,指通过铸造这种工艺将液体浇进铸型里。
二、对本语法的详细阐述
1) 当定语从句的引导词指代事物且介词放在关系代词的前面时,关系代词只能用 which,不能用 that,并且不能省略。如注释一中的 by which。
　e.g. I can't remember the age at which he won the prize.
　　　我不记得他得奖的年龄了。
2) 语法作用:"介词 + 关系代词"在定语从句中充当状语:表示地点,时间和原因的"介词 + which"分别相当于 where, when, why。
　e.g. The earth on which / where we live is a planet.
　　　我们居住的地球是一个行星。
　e.g. I'll never forget the day on which / when I joined the League.
　　　我永远忘不了我加入社团的那一天。
　e.g. Is there any reason for which / why you should have a holiday?
　　　你有理由放假吗?
3) "of + which"起形容词的作用,相当于 whose(用来指物),其词序通常是"n. + of

# Lesson 1　Casting

which"。

e. g. They live in a house whose door /the door of which opens to the south.

　　他们住在一所房子里,房门朝南开。

e. g. He's written a book whose name /the name of which I've completely forgotten.

　　他写了本书,书名我完全忘记了。

4) 可引导限制性定语从句和非限制性定语从句

e. g. There is a rocket by which the direction of the satellite can be changed. (限制性定语从句)

　　人造卫星的方向能够通过火箭被改变。

e. g. We carefully studied the photos, in which we could see signs of plant disease. (非限制性定语从句)

　　我们仔细研究了这些照片,从中我们能看出植物疾病的迹象。

5) 习惯搭配中的介词在定语从句中不能省。

e. g. These are the wires with which different machines are connected. (be connected with 是习惯搭配)

　　这些是连接不同机器的电线。

6) 复杂介词须保持其完整形式,如 by means of (通过…方法), at the back of (在…的后面) 等。

e. g. Sound is a tool by means of which people communicate with each other.

　　声音是人们互相交流的工具。

2. When cured, it is then removed from the damp clay, incidentally destroying the fine details in undercuts present in the clay, but which are now captured in the mold.

　　硬化后,再从湿土中取出,有时会破坏泥坯上细小的凸起部分,但可以在模具中得到。

3. Generally, the form takes less than a week to prepare, after which a production rate of 1-10 units/hr-mold is achieved, with items as massive as 45 kg and as small as 30 g with very high surface resolution and fine tolerances.

　　一般来讲,这(石膏模形成)需要不到一个星期的时间做准备,这样它的生产率就在每个模具 1~10 单元/小时,铸件质量大的可达 45 公斤,小的可以到 30g,而且零件表面具有较低的粗糙度和较高精度。

4. Plaster itself may be cast, as can other chemical setting materials such as concrete or plastic resin - either using single-use waste molds as noted above or multiple-use piece molds, or molds made of small ridged pieces or of flexible material such as latex rubber (which is in turn supported by an exterior mold).

石膏本身也可以铸造,同样,其他一些特定成分的化学材料也可以,例如混凝土,塑料树脂,无论是用上面提到的一次性的模具,还是可多次使用的模具,又或者是用小栋木做成的模具,或是用复杂材料例如橡胶做成的铸型(橡胶是靠外部模具依次支撑的)。

5. When casting plaster or concrete, the finished product is, unlike marble, relatively unattractive, lacking in transparency, and so it is usually painted, often in ways that give the appearance of metal or stone.

   在铸造石膏和混凝土时,成品不像大理石,相对没有那么漂亮,透明度不高,所以他们一般需要着色,往往漆成金属或石头的颜色。

6. A simulation of high-quality marble may be made using certain chemically-set plastic resins (for example epoxy or polyester) with powdered stone added for coloration, often with multiple colors worked in.

   高品质的仿真大理石可利用含有特定化学成分的塑料树脂(譬如环氧或聚酯)与粉石混合一起着色,经常会加入多种色彩。

7. From 5000 years ago, when bees wax formed the pattern, to today's high technology waxes, refractory materials and specialist alloys, the castings ensure high-quality components are produced with the key benefits of accuracy, repeatability, versatility and integrity.

   从5000年前的蜂蜡形成铸型,到今天的高科技蜡、耐火材料、专用合金,这种铸件确保了高品质的零部件,这些部件都具有一些关键优点——精确性、可重复性、通用性和完整性。

8. Although generally used for small castings, this process has been used to produce complete aircraft door frames, with steel castings of up to 300 kg and aluminium castings of up to 30 kg.

   尽管这种工艺一般用来进行小件铸造,但现在也完全可以用来生产飞机门框、300kg的钢铸件和30kg的铝铸件。

## *Check your understanding*

Ⅰ. Give brief answers to the following questions.
   1. What is casting?
   2. Which two distinct subgroups is the casting process subdivided into?
   3. Compared to permanent mold casting, what features does sand casting have?
   4. What fields is shell molding applied in?

Ⅱ. Match the items listed in the following two columns.

# Lesson 1　Casting

| | |
|---|---|
| casting | 石膏 |
| gypsum | 黑色金属 |
| investment casting | 部件 |
| mold | 铸造 |
| ferrous material | 熔模铸造 |
| alloy | 铸型 |
| plaster casting | 齿轮 |
| component | 合金 |
| cavity | 石膏模铸造 |
| gear | 腔 |

## 铸　造

（一般而言，）铸造是将液体浇进铸型里，铸型是空心的，然后凝固成形得到预定形状的一种制造工艺过程。铸造工艺的最后一步是顶出固态铸件或打破铸模获得铸件。铸造可用于形成炽热的液态金属或各种合金材料，形成混合材料时，是先将例如环氧树脂，混凝土，石膏，粘土混合，再冷却凝固得到铸件。铸造常用于制造一些复杂的形状，或利用其他方法制造是困难的或是不经济的场合。

铸造是具有6000年历史的古老工艺。现存最古老的铸件是公元前3200年的铜青蛙。

铸造工艺可以分为截然不同的两种：消失模铸造和非消失性模铸造。

图2.1-1　铸件

### 消失模铸造

消失模铸造是一种通用的铸造方法，包括砂模、塑料模、壳模、石膏模和熔模（脱蜡法）。这种铸造方式使用的是一次性的，不可重复使用的模具。

### 消失性石膏模

耐用的石膏中间体一般用于生产青铜雕塑，或是在创作石刻作品时做导向作用。石膏制作完成后，与泥土原料相比，更持久耐用（室内储存条件下），泥土原料为避免开裂必须保持湿润。由于目前石膏的成本低，价格昂贵的青铜铸造和石刻可以延期至资助人出现再着手进行。石膏铸造被称为是一种技术，而非艺术过程，它存在的寿命甚至可以比艺术家还长。

在消失石膏模中，一种简单而薄的石膏模，先用剑麻和粗麻布进行强化，再浇在原始的泥土混合物上面。硬化后，再从湿土中取出，有时会破坏泥坯上细小的凸起部分，但可以在模具中得到。这种模具在稍后的任何时期内（但是只能用一次）用来铸造石膏正像，和原始的黏土一样。这种"石膏"表面可以进一步完善，着色、打蜡后就近似于青铜铸件了。

**砂型铸造**

图2.1-2　砂型铸型中的铸铁

砂型铸造是最流行、最简单的制造方法之一，至今已使用数百年了。砂型铸造与金属型铸造相比，适用于小批量、低成本生产。砂型铸造不仅仅使制造商能够生产低成本的产品，而且它还有其他的优点，譬如说工作范围小。铸件小到手掌大小，大到火车底盘（一种铸件可以产生整个一辆机动轨道车的底盘）都可以用砂型铸造来完成，砂型铸造可以通过不同种类砂子的模具铸造大多数金属。

砂型铸造要求在几天的生产周期的指定时间里输出率要高（每个模具1～20件/小时），它是无与伦比的大型零件的生产方式。湿砂型几乎没有零件重量限制，而干砂型有零件重量的限制，大至2300～2700公斤，小至0.075～0.1公斤。砂子是利用粘土（如湿砂）或化学黏合剂或聚合油（如机油）接合在一起的。砂子在大多数操作中都可以多次回收，额外投资很少。

**石膏模铸造**

石膏模铸造类似于砂型铸造，只是用石膏替代砂子。石膏混合物的组成是70%～80%的石膏和20%～30%的增强剂和水。一般来讲，制备模具需要不到一周的时间，之后生产率就在每个模具1～10单元/小时，铸件质量大的可达45公斤，小的可以到30g，而且零件表面具有较低的粗糙度和较高精度。石膏铸造出的典型零件有锁零件、齿轮、阀门、日用器具、模具和装饰品。由于石膏和模具制造成本低，石膏铸造是一种替代其他铸造工艺的廉价铸造工艺。但是，它也有缺点，模具的质量取决于诸多因素，包括石膏模具成分的浓度、浇注步骤和石膏凝固技术。如果这些因素没有精确检测，那模具将产生变形、烘干后收缩以及模具表面很粗糙。

一旦石膏用过并破裂后，就很难再铸造。石膏铸造通常用于有色金属，如铝、锌或铜合金，它不能用于铸造黑色金属材料，因为石膏里的硫和铁缓慢反应。石膏本身经受不起1200℃以上的高温，这也限制了在石膏模具中铸造金属的数量。在铸型准备好之前，型芯里用脱模剂喷涂一层薄膜来防止铸型与型芯粘在一起。铸型在振动，因此石膏填充了型芯周围的细小孔。一般在15分钟之内,石膏定型而型芯就可以取出了。再将石膏在温度120℃~260℃之间烘干,然后将铸型预热,这样就可以将铸造金属注入了。

石膏铸造象征了铸造工艺复杂性的增加,它的技术含量较高,尽管现在机器人简单地实现了自动化功能,但是这样高精度的型芯的设计工作还需要掌握更高技术的技术人员的直接操作。

**铸造石膏、混凝土和塑料树脂**

石膏本身也可以铸造,同样,其他一些特定成分的化学材料也可以,例如混凝土、塑料树脂,无论是用上面提到的一次性的模具,还是可多次使用的模具,又或者是用小栋木做成的模具,或是用韧性材料例如橡胶做成的铸型(橡胶是靠外部模具支撑的)。在铸造石膏和混凝土时,成品不像大理石,相对没有那么漂亮,透明度不高,所以他们一般需要着色,往往漆成金属或石头的颜色。另外,为了赋予铸件石头一样的色彩成分,最初的几层铸造层可以含有有色砂。铸造混凝土时,不像石膏,它一般用来制作雕塑、喷泉或者户外的石凳。高品质的仿真大理石可利用含有特定化学成分的塑料树脂(譬如环氧或聚酯)与石粉混合一起着色,经常会加入多种色彩。后者则是制作各种漂亮的盥洗台,盥洗台的台面以及淋浴棚的常用方法。随着着色技术的熟练,产生的这些模拟色,与天然大理石和石灰中的颜色已相差无几了。

**壳模**

壳模也类似于砂型,不同的是它利用砂和3%~6%的树脂的混合物将颗粒结合在一起。壳模也用砂,但颗粒比湿砂要小得多。准备和生产壳模铸型需要几个星期时间,输出率将达到每个模具5~50件/小时。铝、镁产品一般重量限制在平均13.5kg,但是壳模铸造也能得到45~90kg的铝镁产品。壳模的壁厚在3~10mm,这完全取决于树脂的形成时间。

壳模适用于精度要求高的小零件,例如齿轮外壳、气缸盖和连杆。它同样也可用来制造高精度的铸模砂心。壳模铸造工艺复杂,使得零件铸造所需劳动力较少。

壳模处理可以分为12个不同步骤:

1. 准备金属模板
2. 混合树脂和砂
3. 预热模板,温度一般在505~550K
4. 翻转模板(砂子在翻斗的一端,模板在翻斗的另一端,翻斗倒置的时间长短取决于壳型的厚度需要)
5. 成型、烘干型壳
6. 移走盖子

7. 得到壳芯

8. 另一半重复以上操作

9. 合型

10. 浇注

11. 取出铸件

12. 清洁处理

树脂可以高温蒸发掉,这样砂和树脂的混合物可以循环利用。

**熔模铸造**

熔模铸造(在艺术界常被成为脱蜡铸造)是具有几千年历史的工艺,脱蜡工艺是最古老的金属成形技术之一。从5000年前的蜂蜡形成铸型,到今天的高科技蜡、耐火材料、专用合金,这种铸件确保了高品质的零部件,这些部件都具有一些关键优点——精确性、可重复性、通用性和完整性。

图 2.1-3 熔模铸造的核电站阀

之所以称为熔模铸造,是因为铸型是被耐火材料附着或者包围。这种蜡铸型需要特别注意,它们不足以承受模具形成过程中的压力。熔模铸造的优点是蜡可以反复使用。

这种工艺适用于用各种不同的金属和高性能合金重复生产成品零件。尽管这种工艺一般用来进行小件铸造,但现在也完全可以用来生产飞机门框,包括300kg的钢铸件和30kg的铝铸件。比之其他铸造工艺,例如压力铸造或砂型铸造,熔模铸造成本较高,但是用熔融铸造出的零件可以有复杂的轮廓,在大多数情况下,铸造出的零件已接近成品,很少需要甚至不需要机械加工。

# Lesson 2  Forging

*1*  Forging is the term for shaping metal by using localized compressive forces. Cold forging is done at room temperature or near room temperature. Hot forging is done at a high temperature, which makes metal easier to shape and less likely to fracture. Warm forging is done at intermediate temperature between room temperature and hot forging temperature. Forged parts can range in weight from less than a kilogram to 170 metric tons. Forged parts usually require further processing to achieve a finished part.

## History

*2*  Forging is one of the oldest known metalworking processes.

*3*  Forging was done historically by a smith using hammer and anvil, and though the use of water power in the production and working of iron dates back to the 12th century, the hammer and anvil are not obsolete. The smithy has evolved over centuries to the forge shop with engineered processes, production equipment, tooling, raw materials and products to meet the demands of modern industry.

*4*  In modern times, industrial forging is done either with presses or with hammers powered by compressed air, electricity, hydraulics or steam. These hammers are large, having reciprocating weight in the thousands of pounds. Smaller power hammers, which have 500 lb (230 kg) or less reciprocating weight, and hydraulic presses are common in art smithies as well. Steam hammers are becoming obsolete.

## Advantages and disadvantages

*5*  Forging results in metal that is stronger than cast or machined metal parts. This stems from the grain flow caused through forging. As the metal is pounded the grains deform to follow the shape of the part, thus the grains are unbroken throughout the part. Some modern parts take advantage of this for a high strength-to-weight ratio.

*6*  Many metals are forged cold, but iron and its alloys are almost always forged hot. This is for two reasons: first, if work hardening were allowed to progress, hard materials such as iron and steel would become extremely difficult to work with; secondly, steel can be strengthened by other means than cold-working, thus it is more economical to hot forge and then heat treat. Alloys that are amenable to precipitation hardening, such as most alloys of aluminium and titanium, can also be hot forged then hardened. Other materials must be strengthened by the forging process itself.

**Hot forging**

7   Hot forging is defined as working a metal above its recrystallization temperature. The main advantage of hot forging is that as the metal is deformed the strain-hardening effects are negated by the recrystallization process.

**Cold forging**

8   Cold forging is defined as working a metal below its recrystallization temperature, but usually around room temperature. If the temperature is above 0.3 times the melting temperature (on an absolute scale) then it qualifies as warm forging.

**Processes**

9   There are many different kinds of forging processes available, however they can be grouped into three main classes:
10   1. Drawn out: length increases, cross-section decreases
11   2. Upset: Length decreases, cross-section increases
12   3. Squeezed in closed compression dies: produces multidirectional flow
13   Common forging processes include: roll forging, swaging, cogging, open-die forging, impression-die forging, press forging, automatic hot forging and upsetting.

**Equipment**

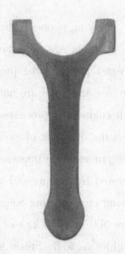

Fig. 2.2 – 1  **A cross-section of a forged connecting rod that has been etched to show the grain flow**

14   The most common type of forging equipment is the hammer and anvil. Principles behind the hammer and anvil are still used today in drop-hammer equipment. The principle behind the machine is very simple--raise the hammer and then drop it or propel it into the workpiece, which rests on the anvil. The main variations between drop-hammers are in the way the hammer is powered; the most common being air and steam hammers. Drop-hammers usually operate in a vertical position. The main reason for this is excess energy (energy that isn't used to deform the workpiece) that isn't released as heat or sound needs to be transmitted to the foundation. Moreover, a large machine base is needed to absorb the impacts.

Lesson 2  Forging

15    To overcome some of the shortcomings of the drop-hammer, the counterblow machine or impactor is used. In a counterblow machine both the hammer and anvil move and the workpiece is held between them. Here excess energy becomes recoil. This allows the machine to work horizontally and consist of a smaller base. Other advantages include less noise, heat and vibration. It also produces a distinctly different flow pattern. Both of these machines can be used for open die or closed die forging.

Fig. 2.2-2  **Hydraulic drop-hammer**
(a) **Material flow of a conventionally forged disc**; (b) **Material flow of an impactor forged disc.**

16    A forging press, often just called a press, is used for press forging. There are two main types: mechanical and hydraulic presses. Mechanical presses function by using cams, cranks or toggles to produce a preset (a predetermined force at a certain location in the stroke) and reproducible stroke. Due to the nature of this type of system different forces are available at different stroke positions. Mechanical presses are faster than their hydraulic counterparts (up to 50 strokes per minute). Their capacities range from 3 to 160 MN (300 to 18,000 tons). Hydraulic presses use fluid pressure and a piston to generate force. The advantages of a hydraulic press over a mechanical press are its flexibility and greater capacity. The disadvantages are that it is slower, larger, and more costly to operate.

17    The roll forging, upsetting, and automatic hot forging processes all use specialized machinery.

### ❖ New Words and Phrases

| | |
|---|---|
| forging ['fɔːdʒiŋ] n. | 锻造 |
| cold forging | 冷锻 |
| hot forging | 热锻 |
| fracture ['fræktʃə] vt. & vi. | 折断 |
| intermediate [ˌintə'miːdjət] adj. | 中间的 |
| metalworking ['metəlˌwɜːkiŋ] n. | 金属加工 |
| smith [smiθ] n. | 铁匠,锻工 |
| hammer ['hæmə] n. | 铁锤 |
| anvil ['ænvil] n. | 铁砧 |
| date back to | 追溯到 |
| obsolete ['ɔbsəliːt] adj. | 过时的 |

| | |
|---|---|
| compress [kəm'pres] vt. | 压缩 |
| hydraulics [hai'drɔːliks] n. | 液压 |
| reciprocate [ri'siprəkeit] vi. | 直线往复运动 |
| cast [kɑːst] n. | 铸造 |
| stem from | 由……造成 |
| pound [paund] vt. & vi. | 连续重击 |
| grain [grein] n. | 晶粒 |
| alloy ['ælɔi] n. | 合金 |
| work hardening | 加工硬化 |
| amenable [ə'miːnəbəl] adj. | 易控制的 |
| precipitation [priˌsipi'teiʃən] n. | 沉淀 |
| precipitation hardening | 淀积硬化 |
| aluminium [ˌælju'miniəm] n. | 铝 |
| titanium [tai'teinjəm] n. | 钛 |
| recrystallization [riˌkristəlai'zeiʃən] n. | 再结晶 |
| negate [ni'geit] vt. | 否定 |
| cross-section n. | 横断截面 |
| rod [rɔd] n. | 杆，棒 |
| connecting rod | 连杆 |
| draw out | 拔长 |
| upset [ʌp'set] vt. & vi. | 镦粗 |
| squeeze [skwiːz] vt. & vi. | 挤压 |
| die [dai] n. | 模 |
| roll forging | 辊锻 |
| swaging ['sweidʒiŋ] n. | 模锻 |
| cogging ['kɔgiŋ] n. | 开坯 |
| open-die forging | 自由锻 |
| press forging | 压锻 |
| impression-die forging | 开式模锻 |
| drop-hammer n. | 落锤 |
| workpiece ['wəːkpiːs] n. | 工件 |
| air hammer | 空气锤 |
| steam hammer | 蒸汽锤 |
| counterblow ['kauntəˌbləu] n. | 对击 |
| horizontally [ˌhɔri'zɔntli] adv. | 水平地 |

# Lesson 2　Forging

| | |
|---|---|
| closed die forging | 闭式模锻 |
| forging press | 锻压机 |
| mechanical press | 机械压力机 |
| hydraulic press | 液压机 |
| cam [kæm] n. | 凸轮 |
| crank [kræŋk] n. | 曲柄 |
| toggle ['tɔgl] n. | 拨动开关 |

## ❖ Notes

1. Forging was done historically by a smith using hammer and anvil, and though the use of water power in the production and working of iron dates to the 12th century, the hammer and anvil are not obsolete.

   锻造是最古老的金属加工工艺之一。早期的锻造是铁匠使用铁锤和铁砧。到了12世纪,出现了利用水压力来生产和加工铁器的技术,尽管这样,铁锤和铁砧依然没有过时。

2. The main advantage of hot forging is that as the metal is deformed the strain-hardening effects are negated by the recrystallization process.

   热锻的主要优点是在再结晶过程中不会发生应变硬化效应,便于金属成形。

3. The main reason for this is excess energy (energy that isn't used to deform the workpiece) that isn't released as heat or sound needs to be transmitted to the foundation.

   这(落锤安装在垂直位置)主要是因为落锤在工作时有很大的冲击力,这种冲击力不能用来成形工件,也不能像热量和噪声一样释放掉,所以就必须转化到地基。

4. Mechanical presses function by using cams, cranks or toggles to produce a preset (a predetermined force at a certain location in the stroke) and reproducible stroke.

   机械压力机利用凸轮、曲柄或拨动开关来产生预设的打击力并重复这样的动作。

## 语法补充:不定式作状语

一、本语法在注释四中的应用

　　"to produce a preset (a predetermined force at a certain location in the stroke) and reproducible stroke"是个带 to 的不定式,在句中表示目的,做目的状语从句,说明机械压力机利用凸轮、曲柄或拨动开关的目的是什么。

二、对本语法的详细阐述

e. g. Helen had to shout to make herself heard above the sound of the music. (目的状语)

　　海伦不得不大喊才能让自己的声音盖过音乐的声音被人听见。

e. g. He got up early in order to/so as to catch the first bus. （目的状语。so as to 不用于句首）
他为了赶上第一班公交车而早起。

e. g. He hurried to the station only to find that the train had left. （结果状语。only to do 表示出乎意料地发现）
他急急忙忙地赶到车站,结果发现火车已经开走了。

e. g. I'm so sorry to hear about your failure in business. （原因状语,用于形容词后表示喜怒哀乐的原因）
听说你生意失败,我很遗憾。

e. g. He is old enough to go to school. （与 enough 结合作结果状语）
他年纪够大了,可以上学了。

e. g. She is too tired to do the job. （与 too 结合作结果状语）
她太累了,以至于不能做这个工作。

e. g. This book is difficult to understand. （不定式作表语形容词的状语,与句子主语构成逻辑上的动宾关系时,不定式多用主动形式,这是因为人们往往认为形容词后省去了 for sb.）。
这本书很难理解。

e. g. This kind of fish is nice to eat.
这种鱼很好吃。

e. g. English is easy to learn.
英语很容易学。

## Check your understanding

Ⅰ. Give brief answers to the following questions.
1. What may be forging classified into?
2. What is the advantage of hot forging?
3. What do common forging processes include?
4. What is the most common type of forging equipment?

Ⅱ. Match the items listed in the following two columns.

| | |
|---|---|
| forging | 连杆 |
| recrystallization | 液压机 |
| forging press | 金属加工 |
| hydraulic press | 锻造 |
| metalworking | 蒸汽锤 |

| | |
|---|---|
| cogging | 再结晶 |
| roll forging | 凸轮 |
| cam | 开坯 |
| connecting rod | 锻压机 |
| steam hammer | 辊锻 |

# 锻 造

锻造是用局部压力加工金属的一种方法。冷锻是在室温或接近室温的情况下加工金属。热锻是在高温情况下对金属进行加工,高温能使金属易于成型和不易于折断。温锻(温壳锻压)的锻造温度在冷锻温度和热锻温度之间。锻件的质量可以小至一公斤以下,大到170吨。锻件一般还需要进一步加工处理才能得到成品。

## 历史

锻造是最古老的金属加工工艺之一。

早期的锻造是铁匠使用铁锤和铁砧。到了12世纪,出现了利用水压力来生产和加工铁器的技术,尽管这样,铁锤和铁砧依然没有过时。铁匠铺经过几个世纪的演变,现在成为拥有加工工艺流程、生产设备、模具、原料和成品以满足现代工业需求的锻造车间。

现在,工业锻造一般利用锻压机械进行锻压,或者利用压缩空气、电机、液压或蒸汽驱动铁锤进行锻压,这种铁锤的质量很大,往复运动质量可达数千磅。小的动力锤往复运动质量也有500磅(230公斤)或者再小一些,这种小的液压锤一般在制造艺术品的铁匠铺比较多见。蒸汽锤已经慢慢被淘汰了。

## 优点和缺点

由于锻造过程中金属晶粒发生移位,金属坯料经锻造后的锻件性能优于铸件或机械加工金属零件的机械性能。金属遭到连续重击后,晶粒发生变形,从而达到设计要求达到的形状,而且在整个过程中晶粒并没有破坏。现在一些零件就利用这一优点来达到高的强度-重量比。

许多金属都可以冷锻,但铁及其合金大多数情况下都是进行热锻。这主要有两个原因:第一,如果进行加工硬化,像钢和铁这样的硬质材料将无法工作。第二,除了冷加工,钢铁的强化可以通过其他方法,因此通过热锻再热处理来得更经济。易于硬化的合金,如大多数的铝合金、钛合金,也可以先热锻再硬化。其他材料必须通过锻造处理来强化。

### 热锻

热锻,锻造温度高于坯料金属的再结晶温度,它的主要优点是在再结晶过程中不会发生应变硬化效应,便于金属成形。

### 冷锻

冷锻,锻造温度低于坯料金属的再结晶温度,一般在常温下。如果温度高于0.3倍的熔融温度(在绝对比例上),则定性为温锻。

### 工艺

不同的锻造方法有不同的锻造工艺,但他们可以分为三种主要类别:

1. 拔长:长度的增加,横断截面的减小;
2. 镦粗:长度的缩短,横断截面的增大;
3. 在封闭挤压模中的挤压:产生多向金属流。

一般锻造工艺包括:辊锻备坯、模锻成形、锻造开坯、自由锻、开式模锻、锻压、自动化热锻和镦粗。

### 设备

最常见的锻压设备是铁锤和铁砧。尽管现在用的是落锤设备,但它的基本原理和铁锤、铁砧的一样非常简单,就是升高锤再落锤,或者将铁锤推进工件,工件置于铁砧上。落锤的主要差别在于铁锤的驱动方式不一样,最常见的是空气锤和蒸汽锤。落锤一般安装在垂直位置,这主要是因为落锤在工作时有很大的冲击力,这种冲击力不能用来成形工件,也不能像热量和噪声一样释放掉,所以就必须转化到地基,并且还需要庞大的机座来吸收振动。

图 2.2-1 锻造连杆横断截面的晶粒变形

为了克服落锤的一些缺点,可以使用对击锤或冲击锤。对击锤中铁锤和铁砧同时移动,工件放在他们之间,这样振动就会相互抵消。这种设备可以水平工作,底座小,噪音低,振动小,散热好,而且它产生不同的物质流态形式。这两种设备均可用于自由锻和开式模锻中。

图 2.2-2 液压落锤设备

锻压机,又叫压力机,是用于锻压的。主要有两种类型:机械压力机和液压机。机械压力机利用凸轮、曲柄或拨动开关来产生预设的打击力并重复这样的动作,它可以在不同的打击位置用不同大小的打击力。机械压力机的打击速度比液压机快(可达每分钟50次),目前有300~18000吨的机械压力机。液压机利用液体压力和活塞来产生打击力。与机械压力机比较,它的优点是灵活,容量大。它的缺点是速度慢,体积大,成本高。

辊锻、镦粗和自动化热锻流程一般使用特殊机器。

# Lesson 3  Welding

*1*  Welding is a fabrication or sculptural process that joins materials, usually metals or thermoplastics, by causing coalescence. This is often done by melting the workpieces and adding a filler material to form a pool of molten material (the weld puddle) that cools to become a strong joint, with pressure created by heat, or by itself, to produce the weld. This is in contrast with soldering and brazing, which involve melting a lower-melting-point material between the workpieces to form a bond between them, without melting the workpieces.

*2*  Many different energy sources can be used for welding, including a gas flame, an electric arc, a laser, an electron beam, friction, and ultrasound. However often as an industrial process, welding can be done in many different environments, including open air, under water and in outer space. Regardless of location, however, welding remains dangerous, and precautions must be taken to avoid burns, electric shock, eye damage, poisonous fumes, and overexposure to ultraviolet light.

Fig. 2.3 – 1  **Arc welding**

*3*  Until the end of the 19th century, the only welding process was forge welding, which blacksmiths had used for centuries to join metals by heating and pounding them. Arc welding and oxyfuel welding were among the first processes to develop late in the century, and resistance welding followed soon after. Welding technology advanced quickly during the early 20th century as World War I and World War II drove the demand for reliable and inexpensive joining methods. Following the wars, several modern welding techniques were developed, including manual methods like shielded metal arc welding, now one of the most popular welding methods, as well as semi-automatic and automatic processes such as gas metal arc welding, submerged arc welding, flux-cored arc welding and electroslag welding. Developments continued with the invention of laser beam welding and electron beam welding in the latter half of the century. Today, the science continues to advance. Robot welding is becoming more commonplace in industrial settings, and researchers continue to develop new welding methods to gain greater understanding of weld quality and properties.

4  One of the most common types of arc welding is shielded metal arc welding (SMAW), which is also known as manual metal arc welding (MMA) or stick welding. Electric current is used to strike an arc between the base material and consumable electrode rod, which is made of steel and is covered with a flux that protects the weld area from oxidation and contamination caused by $CO_2$ gas generated during the welding process.

Fig. 2.3 – 2  **Shielded metal arc welding**

The electrode core itself acts as filler material, making a separate filler unnecessary.

5  The process is versatile and can be performed with relatively inexpensive equipment, making it well suited to shop jobs and field work. An operator can become reasonably proficient with a modest amount of training and can achieve mastery with experience. Weld speeds are rather slow, since the consumable electrodes must be frequently replaced and because slag, the residue from the flux, must be chipped away after welding. Furthermore, the process is generally limited to welding ferrous materials, though special electrodes have made possible the welding of cast iron, nickel, aluminium, copper, and other metals. Inexperienced operators may find it difficult to make good out-of-position welds with this process.

6  Gas metal arc welding (GMAW), also known as metal inert gas or MIG welding, is a semi-automatic or automatic process that uses a continuous wire feed as an electrode and an inert or semi-inert gas mixture to protect the weld from contamination. As with SMAW, reasonable operator proficiency can be achieved with modest training. Since the electrode is continuous, welding speeds are greater for GMAW than for SMAW. Also, the smaller arc size compared with the shielded metal arc welding process makes it easier to make out-of-position welds (e. g. overhead joints, as would be welded underneath a structure).

7  The equipment required to perform the GMAW process is more complex and expensive than that required for SMAW, and requires a more complex setup procedure. Therefore, GMAW is less portable and versatile, and due to the use of a separate shielding gas, is not particularly suitable for outdoor work. However, owing to the higher average rate at which welds can be completed, GMAW is well suited to production welding. The process can be applied to a wide variety of metals, both ferrous and non-ferrous.

*8*  A related process, flux-cored arc welding (FCAW), uses similar equipment but uses wire consisting of a steel electrode surrounded by a powder fill material. This cored wire is more expensive than the standard solid wire and can generate fumes and/or slag, but it permits even higher welding speed and greater metal penetration.

*9*  Gas tungsten arc welding (GTAW), or tungsten inert gas (TIG) welding (also sometimes erroneously referred to as heliarc welding), is a manual welding process that uses a nonconsumable tungsten electrode, an inert or semi-inert gas mixture, and a separate filler material. Especially useful for welding thin materials, this method is characterized by a stable arc and high quality welds, but it requires significant operator skill and can only be accomplished at relatively low speeds.

*10*  GTAW can be used on nearly all weldable metals, though it is most often applied to stainless steel and light metals. It is often used when quality welds are extremely important, such as in bicycle, aircraft and naval applications. A related process, plasma arc welding, also uses a tungsten electrode but uses plasma gas to make the arc. The arc is more concentrated than the GTAW arc, making transverse control more critical and thus generally restricting the technique to a mechanized process. Because of its stable current, the method can be used on a wider range of material thicknesses than can the GTAW process, and furthermore, it is much faster. It can be applied to all of the same materials as GTAW except magnesium, and automated welding of stainless steel is one important application of the process. A variation of the process is plasma cutting, an efficient steel cutting process.

*11*  Submerged arc welding (SAW) is a high-productivity welding method in which the arc is struck beneath a covering layer of flux. This increases arc quality, since contaminants in the atmosphere are blocked by the flux. The slag that forms on the weld generally comes off by itself, and combined with the use of a continuous wire feed, the weld deposition rate is high. Working conditions are much improved over other arc welding processes, since the flux hides the arc and almost no smoke is produced. The process is commonly used in industry, especially for large products and in the manufacture of welded pressure vessels. Other arc welding processes include atomic hydrogen welding, carbon arc welding, electroslag welding, electrogas welding, and stud arc welding.

## ❖ New Words and Phrases

fabrication [ˌfæbriˈkeiʃən] n.     制作
sculptural [ˈskʌlptʃərəl] a.     雕刻的,雕塑的
thermoplastic [ˌθəːməˈplæstik] n.     热塑性塑料
coalescence [ˌkəuəˈlesns] n.     合并,结合,联合

# Lesson 3　Welding

| | |
|---|---|
| melt [melt] *vt.* & *vi.* | 使融化,使熔解 |
| filler [ˈfilə] *n.* | 填充料,掺入物 |
| conjunction [kənˈdʒʌŋkʃən] *n.* | 结合,联合 |
| arc welding | 电弧焊 |
| electric arc | 电弧 |
| electron [iˈlektrɔn] *n.* | 电子 |
| beam [biːm] *n.* | 光线 |
| electron beam | 电子束 |
| friction [ˈfrikʃən] *n.* | 摩擦 |
| ultrasound [ˈʌltrəˌsaund] *n.* | 超声波 |
| precaution [priˈkɔːʃn] *n.* | 预防措施 |
| poisonous [ˈpɔiznəs] *adj.* | 有毒的 |
| fume [fjuːm] *n.* | 烟雾,气味 |
| overexposure [ˈəuvəriksˈpəuʒə] *n.* | 过度暴露 |
| ultraviolet [ˈʌltrəˈvaiəlit] *adj.* | 紫外光的 |
| forge [fɔːdʒ] *n.* | 熔炉,铁工厂 |
| forge welding | 锻焊 |
| blacksmith [ˈblækˌsmiθ] *n.* | 铁匠,锻工 |
| oxyfuel welding | 气焊 |
| resistance [riˈzistəns] *n.* | 电阻 |
| resistance welding | 电阻焊 |
| shielded [ˈʃiːldid] *adj.* | 有屏蔽的 |
| shielded metal arc welding | 焊条电弧焊 |
| gas metal arc welding | 熔化极气体保护电弧焊 |
| submerged [səbˈməːdʒd] *adj.* | 水下的 |
| submerged arc welding | 埋弧焊 |
| flux [flʌks] *n.* | 流出 |
| cored [kɔːd] *adj.* | 带心的 |
| flux cored arc welding | 药芯焊丝电弧焊 |
| electroslag [iˈlektrəuslæg] *n.* | 电炉渣 |
| electroslag welding | 电渣焊 |
| laser [ˈleizə] *n.* | 激光 |
| laser beam welding | 激光焊 |
| electron beam welding | 电子束焊 |
| oxidation [ɔksiˈdeiʃən] *n.* | 氧化 |

| | | |
|---|---|---|
| contamination [kən͵tæmi'neiʃən] n. | | 污染物 |
| versatile ['və:sətail] adj. | | 多用途的 |
| proficient [prə'fiʃənt] adj. | | 精通的,熟练的 |
| modest ['mɔdist] adj. | | 适度的 |
| mastery ['mɑ:stəri] n. | | 精通,熟练 |
| slag [slæg] n. | | 熔渣 |
| residue ['rezidju:] n. | | 余渣 |
| chip away | | 清除 |
| cast iron | | 铸铁 |
| nickel ['nikl] n. | | 镍 |
| copper ['kɔpə] n. | | 铜 |
| penetration [peni'treiʃən] n. | | 穿过,渗透,突破 |
| significant [sig'nifikənt] adj. | | 有重大意义的 |
| naval ['neivəl] adj. | | 海军的 |
| transverse ['trænzvə:s] adj. | | 横向的 |
| critical ['kritikəl] adj. | | 关键性的 |
| restrict [ris'trikt] vt. | | 限制,约束 |
| productivity [͵prɔdʌk'tiviti] n. | | 生产率 |
| deposition [͵depə'ziʃən] n. | | 沉积作用 |
| deposition rate | | 沉积速度 |
| stud arc welding | | 螺栓电弧焊 |

## ❖ Notes

1. This is often done by melting the workpieces and adding a filler material to form a pool of molten material (the weld puddle) that cools to become a strong joint, with pressure sometimes used in conjunction with heat, or by itself, to produce the weld.

   焊接一般是通过熔解工件和加入一些填充材料,形成一堆熔融材料,冷却后就成了牢固的接头,焊接时常使用热能产生的压力,也可以无需外力来得到焊接件。

**语法补充:动名词的用法(主语、表语、宾语、定语)**

一、本语法在注释一中的应用

　　动名词是指在动词后面加上后缀-ing,使之与名词有同样的语法功能。"by melting the workpieces and adding a filler material"中 melting 和 adding 就是动名词,分别与 the workpieces 和 a filler material 构成两个动名词短语,共同做 by 的宾语。

二、对本语法的详细阐述

# Lesson 3    Welding

e. g. <u>Seeing</u> is believing. （主语）眼见为实。
e. g. My job is <u>teaching English</u>. （表语）我的工作是教英语。
e. g. She is good at <u>playing the piano</u>. （宾语）她擅长弹钢琴。
e. g. Would you mind <u>opening the door</u>? （宾语）你介意开门吗?
e. g. It was so funny that I couldn't help <u>laughing</u>. （宾语）我忍不住笑,这挺滑稽。
e. g. Do you have problems <u>finishing the work</u>? （宾语）你完成这工作有困难吗?
e. g. He dislikes <u>eating</u> in KFC. （宾语）他不喜欢在肯德基吃饭。
e. g. She is in the <u>living</u> room. （定语,表示用途,所属关系）她在起居室。

2. Following the wars, several modern welding techniques were developed, including manual methods like shielded metal arc welding, now one of the most popular welding methods, as well as semi-automatic and automatic processes such as gas metal arc welding, submerged arc welding, flux-cored arc welding and electroslag welding.
战争之后,一些现代焊接工艺技术得到了发展,包括一些手工方法,例如焊条电弧焊,它是目前最流行的焊接方法之一,还有半自动和自动工艺的溶化极气体保护电弧焊、埋弧焊、药芯焊丝电弧焊和电渣焊。

3. Gas tungsten arc welding (GTAW), or tungsten inert gas (TIG) welding (also sometimes erroneously referred to as heliarc welding), is a manual welding process that uses a nonconsumable tungsten electrode, an inert or semi-inert gas mixture, and a separate filler material.
钨极气体保护电弧焊(GTAW),或称钨惰性气体焊(TIG)(有时也被错误的称为氩弧焊),它是一种手工焊接工艺,它使用不可熔化式的钨电极,惰性气体或者是半混合惰性气体混合物及一种单独的填充材料。

## Check your understanding

Ⅰ. Give brief answers to the following questions.
   1. What is welding?
   2. What is the most common type of arc welding ?

Ⅱ. Match the items listed in the following two columns.
   arc welding              埋弧焊
   resistance welding       锻焊
   forge welding            药芯焊丝电弧焊
   submerged arc welding    螺栓电弧焊
   stud arc welding         电弧焊

oxyfuel welding　　　　　　　氧乙炔焊
flux cored arc welding　　　　电阻焊
laser beam welding　　　　　　气焊

# 焊　　接

　　焊接是一种制作和雕刻工艺,它可以使材料,通常是金属或热塑性塑料之间合并连接起来。焊接一般是通过熔解工件和加入一些填充材料,形成一堆熔融材料,冷却后就成了牢固的接头。焊接时常使用热能产生的压力,也可以无需外力来得到焊接件。与此截然不同的是,锡焊和铜焊是通过熔解工件之间的低熔点材料以连接工件,不需要融化工件。

　　许多不同能源都可以用来焊接,包括天然气火焰、电弧、激光、电子束、摩擦和超声波。虽然焊接通常是工业生产流程,有可能在不同的环境下进行,例如户外环境、水下环境或是太空环境。即使不考虑环境,焊接也是很危险的,因此,防护措施必须要有,避免灼伤、电击眼睛损伤、有毒气体、过度暴露在紫外光下。

　　一直到19世纪末,焊接方法只有锻焊,数百年来铁匠一直利用这种工艺通过加热和打击金属来连接金属。电弧焊和气焊是19世纪末最早开发的焊接方法中的两种,然后就有了电阻焊。由于第一次世界大战和第二次世界大战需要产生可靠的、廉

图 2.3-1　电弧焊

价的连接方法,焊接工艺就在20世纪早期得到了迅速发展。战争之后,一些现代焊接工艺技术得到了发展,包括一些手工方法,例如焊条电弧焊,它是目前最流行的焊接方法之一,还有半自动和自动工艺的溶化极气体保护电弧焊、埋弧焊、药芯焊丝电弧焊和电渣焊,激光焊和电子束焊在接下来的半个世纪继续发展。今天,科技依然在进步,在工业生产中焊接机器人的使用越来越普遍,科学家还在继续开发新的焊接方法以获得更好质量和性能的焊接件。

　　最常见的电弧焊之一是焊条电弧焊(SMAW),也称为手工金属电弧焊(MMA)或焊条焊接。电流可以使母材和可熔化性电极棒粘在一起,这种电极棒的材料是钢,表面附有熔剂,它可以防止焊接过程中产生的 $CO_2$ 污染和氧化焊接区。焊条芯本身也作为填充材料,就不需要单独的填充料了。

　　这种工艺可以适用于多种场合,并且设备成本较低,非常适用于工厂作业和野外

作业。操作者通过适量的培训即可熟练操作,并通过一段时间的锻炼而精通。焊接速度相当缓慢,这主要是因为焊条必须经常更换,而且熔剂产生的熔渣必须要在焊后清除。此外,这种焊接工艺不适于焊接有色金属,只有采用特殊电极时才有可能焊接铸铁、镍、铝、铜以及其他金属。不熟练的操作工就很难利用这种工艺做出位置不寻常的优质焊件。

图 2.3-2 焊条电弧焊

熔化极气体保护电弧焊,也被称为金属惰性气体焊,简称 MIG 焊接,这是一种半自动或自动的焊接方法,它的电极利用连续送丝,使用惰性气体和半混合惰性气体混合物来防止焊件被污染。和焊条电弧焊一样,一名合格操作工可以通过适当的培训来熟练工作。由于熔化极气体保护电弧焊的电极是连续进给的,焊接速度大于焊条电弧焊的焊接速度。此外,由于此焊法的弧度尺寸比焊条电弧焊小,所以它可以更容易制造出位置不寻常的焊接件(例如空间接头,它可以在建筑物下面实现焊接)。

熔化极气体保护电弧焊所需的设备比焊条电弧焊所需的更贵,更复杂,需要更复杂的安装程序。这种焊接工艺不够便携,使用范围也有限,这主要因为它需要使用单独的保护气体,不适合户外工作。尽管这样,熔化极气体保护焊的平均焊接速率较高,非常适用于生产焊接,并可应用于各种金属,包括有色金属和黑色金属。

前面提到的药芯焊丝电弧焊(简称 FCAW)使用了类似的设备,但使用钢丝组成的焊条,里面有粉末填充材料,这种药芯焊条比普通焊条昂贵,且易产生烟雾或者熔渣,或两者皆有,但它的焊接速度更高,具有更大的金属穿透力。

钨极气体保护电弧焊(GTAW),或称钨惰性气体焊(TIG)(有时也被错误地称为氩弧焊),是一种手工焊接工艺,它使用不可熔化式的钨电极,惰性气体或者是半混合惰性气体混合物及一种单独的填充材料。这种焊接非常适用于薄件焊接,它的优点就是拥有稳定的电弧和高质量的焊件。但是,它需要较高的操作技巧,焊接速度也相对较慢。

钨极气体保护电弧焊适用于几乎所有可焊接的金属,但一般用于不锈钢和轻金属。它经常用于高质量的焊件特别重要的场合,例如自行车、飞机以及海军上应用较多。上面提到的等离子电弧焊,也使用钨电极,不同的是它使用等离子气体产生电弧。这种电弧比 GTAW 的电弧更集中,使横向控制更为重要,这样使焊接技术更趋于机械化工艺。由于其稳定的电流,这种焊接方法可用于比 GTAW 更广泛的材料厚度,此外,它的焊接速度快得多。这种方法可以用到所有 GTAW 所能焊接除镁之外的金属中,一个重要的应用就是不锈钢的自动焊接。这种的工艺的变体是等离子切割,是一种高效率的钢切割过程。

埋弧焊是一种高效的焊接方法,在这种焊接方法中电弧是被粘于熔剂的覆盖层之下,而熔剂阻止了大气中的污染物进入,这样就强化了电弧质量。在焊接过程中形成的熔渣一般会自动脱落下来,再加上使用连续送丝,焊接沉积速度高。与其他的电弧焊工艺相比,这种焊接工艺大大改善了工作条件,这是因为电弧藏于熔剂之下,几乎没有烟雾产生。这种工艺常用于工业之中,尤其适用于生产大型产品以及制造一些焊接压力容器。还有其他的一些电弧焊,包括原子氢焊、碳弧焊、电渣焊、气电立焊以及螺栓电弧焊。

**构词法之派生法**
派生法的构成方式(续)
(二) 后缀
1. 构成名词的后缀
1) -tion, -ing, -ment, -ant, -ture, -ness, -ity
e. g. application : [名词] 应用
　　　understanding：[名词] 理解
　　　development：[名词] 发展
　　　determinant：[名词] 决定因素
　　　mixture：[名词] 混合物
　　　thickness：[名词] 厚度
　　　versatility：[名词] 多功能
2) -graphy：表示学科和研究
e. g. crystallography：[名词] 结晶学
3) -er, -or：表示做某事的人或物
e. g. strengthener：[名词] 增强剂
　　　operator：[名词] 操作工
4) -ist：表示"……的实行者"
e. g. metallurgist：[名词] 冶金家
2. 构成动词的后缀
-en, -fy, -ize, -ate
e. g. broaden：[动词] 加宽
　　　classify：[动词] 分类
　　　localize：[动词] 使局部化
　　　necessitate：[动词] 使……成为必需

# Unit 3

## Lesson 1  Lathe (metal)

*1*  Metal lathe or metalworking lathe are generic terms for any of a large class of lathes designed for precisely machining relatively hard materials. They were originally designed to machine metals; however, with the advent of plastics and other materials, and with their inherent versatility, they are used in a wide range of applications, and a broad range of materials. In machining jargon, where the larger context is already understood, they are usually simply called lathes, or else referred to by more-specific subtype names (toolroom lathe, turret lathe, etc.). These rigid machine tools remove material from a rotating workpiece via the (typically linear) movements of various cutting tools, such as tool bits and drill bits.

Fig. 3.1 – 1  **an example of a lathe**

### Construction

*2*  The machine has been greatly modified for various applications, however, a familiarity with the basics shows the similarities between types. These machines consist of, at the least, a headstock, bed, carriage and tailstock. The better machines are solidly constructed with broad bearing surfaces (slides or ways) for stability and manufactured with greater precision. This helps ensure the components manufactured on the machines can meet the required tolerances and repeatability.

**Headstock**

Fig. 3.1 - 2  **Basic components of headstock**

3  The headstock (H1) houses the main spindle (H4), speed change mechanism (H2, H3), and change gears (H10). The headstock is required to be made as robust as possible due to the cutting forces involved, which can distort a lightly built housing, and induce harmonic vibrations that will transfer through to the workpiece, reducing the quality of the finished workpiece.

4  The main spindle is generally hollow to allow long bars to extend through to the work area, which reduces preparation and waste of material. The spindle then runs in precision bearings and is fitted with some means of attaching work holding devices such as chucks or faceplates. This end of the spindle will also have an included taper, usually morse, to allow the insertion of tapers and centers. On older machines the spindle was directly driven by a flat belt pulley with the lower speeds available by manipulating the bull gear while later machines use a gear box driven by a dedicated electric motor. The fully geared head allows the speed selection to be done entirely through the gearbox. [2]

**Bed**

5  The bed is a robust base that connects to the headstock and permits the carriage and tailstock to be aligned parallel with the axis of the spindle. This is facilitated by hardened and ground ways which restrain the carriage and tailstock in a set track. The carriage travels

# Lesson 1  Lathe (metal)

by means of a rack and pinion system, leadscrew of accurate pitch, or feed screw.

**Feed and lead screws**

6  The feed screw (H8) is a long driveshaft that allows a series of gears to drive the carriage mechanisms. These gears are located in the apron of the carriage. Both the feed screw and leadscrew (H9) are driven by either the change gears (on the quadrant) or an intermediate gearbox known as a quick change gearbox (H6) or Norton gearbox. These intermediate gears allow the correct ratio and direction to be set for cutting threads or worm gears. Tumbler gears (operated by H5) are provided between the spindle and gear train along with a quadrant plate that enables a gear train of the correct ratio and direction to be introduced. This provides a constant relationship between the number of turns the spindle makes, with the number of turns the leadscrew makes. This ratio allows screwthreads to be cut on the workpiece without the aid of a die.

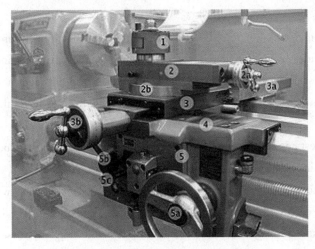

Fig. 3.1 – 3  **Basic components of carriage**

7  The leadscrew will be manufactured to either imperial or metric standards and will require a conversion ratio to be introduced to create thread forms from a different family. To accurately convert from one thread form to the other requires a 127-tooth gear, or on lathes not large enough to mount one, an approximation may be used. Multiples of 3 and 7 giving a ratio of 63:1 can be used to cut fairly loose threads. This conversion ratio is often built into the quick change gearboxes.

**Carriage**

8  In its simplest form the carriage holds the tool bit and moves it longitudinally (turning) or perpendicularly (facing) under the control of the operator. The operator moves the

carriage manually via the handwheel (5a) or automatically by engaging the feedscrew with the carriage feed mechanism (5c), which provides some relief for the operator as the movement of the carriage becomes power assisted. The handwheels (2a, 3b, 5a) on the carriage and its related slides are usually calibrated, both for ease of use and to assist in making reproducible cuts.

### Cross-slide

9  The cross-slide stands atop the carriage and has a leadscrew that travels perpendicular to the main spindle axis, which permits facing operations to be performed. This leadscrew can be engaged with the feedscrew (mentioned previously) for providing automated movement to the cross-slide. Only one direction can be engaged at a time as an interlock mechanism will shut out the second gear train.

### Compound rest

10  The compound rest (or top slide) is the part of the machine where the tool post is mounted. It provides a smaller amount of movement along its axis via another leadscrew. The compound rest axis can be adjusted independently of the carriage or cross-slide. It is utilized when turning tapers, when screw cutting or to obtain finer feeds than the leadscrew normally permits.

11  The slide rest can be traced to the fifteenth century. The story has long circulated that Henry Maudslay invented it, but he did not (and never claimed so). The legend that Maudslay invented the slide rest originated with James Nasmyth, who wrote ambiguously about it in his Remarks on the Introduction of the Slide Principle, 1841; later other writers misunderstood, and propagated the error. Maudslay did help to disseminate the idea widely. It is highly probable that he saw it when he was working at the Arsenal as a boy. In 1794, whilst he was working for Joseph Bramah, he made one, and when he had his own workshop used it extensively in the lathes he made and sold them. Coupled with the network of engineers he trained, this ensured the slide rest became widely known and copied by other lathe makers, and so diffused throughout British engineering workshops. A practical and versatile screw-cutting lathe incorporating the trio of leadscrew, change gears, and slide rest was Maudslay's most important achievement.

12  The first fully documented, all-metal slide rest lathe was invented by Jacques de Vaucanson around 1751. It was described in the Encyclopedia a long time before Maudslay invented and perfected his version. It is likely that Maudslay was not aware of Vaucanson's work, since his first versions of the slide rest had many errors which were not present in the

Vaucanson lathe.

**Toolpost**

*13*  The tool bit is mounted in the toolpost which may be of the American lantern style, traditional 4 sided square style, or in a quick change style such as the multifix arrangement pictured. The advantage of a quick change set-up is to allow an unlimited number of tools to be used (up to the number of holders available) rather than being limited to 1 tool with the lantern style, or 3 to 4 tools with the 4 sided type. Interchangeable tool holders allow all the tools to be preset to a center height that will not change, even if the holder is removed from the machine.

**Tailstock**

*14*  The tailstock is a toolholder directly mounted on the spindle axis, opposite the headstock. The spindle (T5) does not rotate but does travel longitudinally under the action of a leadscrew and handwheel (T1). The spindle includes a taper to hold drill bits, centers and other tooling. The tailstock can be positioned along the bed and clamped (T6) in position as required. There is also provision to offset the tailstock (T4) from the spindles axis, which is useful for turning small tapers.

*15*  The image shows a reduction gear box (T2) between the handwheel and spindle, which is a feature found only in the larger center lathes, where large drills may necessitate the extra leverage.

Fig. 3.1 – 4  **Basic components of tailstock**

## ❖ New Words and Phrases

| | | |
|---|---|---|
| generic | [dʒiˈnerik] a. | 一般的,普通的,共有的 |
| originally | [əˈridʒənəli] ad. | 本来,原来,最初,重要的 |
| advent | [ˈædvənt] n. | 出现,到来 |
| inherent | [inˈhiərənt] a. | 内在的,固有的 |
| versatility | [ˌvəːsəˈtiləti] n. | 多才多艺,用途广泛,万能 |
| jargon | [ˈdʒɑːgən] n. | 行话 |
| toolroom | [ˈtuːlruːm] n. | 工具室(工具车间) |
| toolroom lathe | | 工具车床 |
| turret | [ˈtʌrit] n. | 小塔,角楼 |
| turret lathe | | 六角车床 |
| rigid | [ˈridʒid] a. | 僵硬的,刻板的,严格的,刚性的 |
| tool bit | | 刀具,刀头 |
| drill bit | | 钻头 |
| headstock | [ˈhedstɔk] n. | 主轴承,床头箱,主轴箱 |
| carriage | [ˈkæridʒ] n. | 溜板,拖板 |
| tailstock | [ˈteilstɔk] n. | 尾架(顶尖座,滑轮活轴,托柄尾部) |
| bearing surface | | 轴承支承面 |
| slide | [slaid] n. | 滑轨 |
| way | [wei] n. | 滑道,(导)轨 |
| tolerance | [ˈtɔlərəns] n. | 公差 |
| mechanism | [ˈmekənizəm] n. | 机械,机构,结构,机制,原理 |
| robust | [rəˈbʌst] a. | 强壮的,强健的,坚固的,结实的,耐用的 |
| housing | [ˈhauziŋ] n. | 房屋(外壳,外套,外罩,住宅,卡箍,遮盖物) |
| harmonic | [hɑːˈmɔnik] a. | 调和的,音乐般的,和声的;n. 和音,调波 |
| attaching | [əˈtætʃiŋ] a. | 附属的 |
| chuck | [tʃʌk] n. | 卡盘 |
| be fitted with | | 配备 |
| manipulate | [məˈnipjuleit] v. | 操纵,利用,假造 |
| pulley | [ˈpuli] n. | 皮带轮 |
| bull gear | | 大齿轮 |
| aligned | [əˈlaind] a. | 对齐的,均衡的 |
| parallel | [ˈpærəlel] a. | 平行的;ad. 平行地;n. 平行线 |
| facilitate | [fəˈsiliteit] v. | 帮助,使……容易,促进 |
| restrain | [risˈtrein] v. | 抑制,阻止,束缚 |

## Lesson 1    Lathe (metal)

| | |
|---|---|
| rack [ræk] *n.* | 架,行李架,拷问台;齿条 |
| pinion [ˈpinjən] *n.* | 鸟翼,小齿轮 |
| leadscrew [liːdskruː] *n.* | 导杆(推垃螺杆,丝杠) |
| pitch [pitʃ] *n.* | 程度,投掷,音高,节距 |
| feedscrew [fiːdskruː] *n.* | 进给螺杆,螺旋给料机,螺旋送料机 |
| driveshaft [draivʃɑːft] *n.* | 驱动杆;驱动轴,传动轴,主动轴 |
| quadrant [ˈkwɔdrənt] *n.* | 象限 |
| intermediate [ˌintəˈmiːdjət] *a.* | 中级的,中间的;*n.* 中间体,媒介物 |
| tumbler [ˈtʌmblə] *n.* | 不倒翁 |
| tumbler gear | 摆动换向齿轮 |
| quadrant plate | 挂轮板 |
| imperial [imˈpiəriəl] *a.* | 帝国的,英制的 |
| conversion [kənˈvəːʃən] *n.* | 转变,改变信仰,换位 |
| thread form | 牙型 |
| convert from | 改变…… |
| mount [maunt] *v.* | 增长,装上,爬上,乘马,安装 |
| longitudinally [lɔndʒiˈtjuːdinli] *ad.* | 纵 |
| perpendicularly [ˌpəːpənˈdikjuləli] *ad.* | 垂直(笔直,纵) |
| calibrate [ˈkælibreit] *v.* | 校准(刻度,使……标准化,测定) |
| ease [iːz] *n.* | 安乐,安逸,悠闲; |
| *v.* | 使……安乐,使……安心,减轻,放松 |
| reproducible [ˌriːprəˈdjuːsəbl] *a.* | 可再生的,可复写的,能繁殖的 |
| atop [əˈtɔp] *ad.* | 在顶上 |
| be engaged with | 和……有事,从事(某事) |
| compound [ˈkɔmpaund] *n.* | 混合物,复合 |
| compound rest | 复式刀架;小刀架 |
| tool post | 刀架,刀座 |
| adjusted [əˈdʒʌstid] *a.* | 调整过的 |
| fine feed | 微小进给 |
| slide rest | 滑动刀架 |
| ornamental [ˌɔːnəˈmentl] *a.* | 装饰的 |
| Royal Arsenal | 皇家兵工厂 |
| boring [ˈbɔːriŋ] *n.* | 钻孔 |
| ambiguously [æmˈbigjuəsli] *ad.* | 含糊不清地 |
| propagate [ˈprɔpəgeit] *v.* | 繁殖,传播,传送 |

| | | |
|---|---|---|
| propagated error | | 传递误差 |
| diffused [di'fju:zd] a. | | 散布的，普及的，扩散的 |
| incorporate [in'kɔ:pəreit] a. | | 合并的，公司组织的，具体化的； |
| | v. | 合并，组成公司 |
| encyclopedia [en͵saiklou'pi:diə] n. | | 百科全书 |
| provision [prə'viʒən] n. | | 规定，条款；准备，食物，供应品 |
| offset ['ɔ:fset] n. | | 抵消，支派，位移，偏移；v. 弥补，抵消 |
| necessitate [ni'sesiteit] v. | | 迫使，使……成为必需，需要 |

## ❖ Notes

1. These rigid machine tools remove material from a rotating workpiece via the (typically linear) movements of various cutting tools, such as tool bits and drill bits.
   这些刚性机床通过做直线运动的各种切削刀具(如车刀或钻头)来切削旋转运动的工件上的多余的材料。
2. The fully geared head allows the speed selection to be done entirely through the gearbox.
   通过齿轮箱使这种动力头的速度选择进行得更加彻底。
3. The bed is a robust base that connects to the headstock and permits the carriage and tailstock to be aligned parallel with the axis of the spindle。
   结实的床身连接主轴箱并使拖板箱和尾座沿着床身上的导轨作平形于主轴方向的直线运动。
4. The carriage travels by means of a rack and pinion system, leadscrew of accurate pitch, or feedscrew.
   拖板箱由齿轮齿条机构、准确螺距的丝杠或光杠带动。
5. Both the feedscrew and leadscrew (H9) are driven by either the change gears (on the quadrant) or an intermediate gearbox known as a quick change gearbox (H6) or Norton gearbox.
   光杠和丝杠由四分仪上的变速齿轮、中间齿轮箱即快速进给齿轮箱或诺顿齿轮箱驱动。
6. These intermediate gears allow the correct ratio and direction to be set for cutting threads or worm gears.
   这些中间齿轮通过改变主轴和丝杠的速比来确定螺距和旋向来加工螺杆和蜗杆。
7. Tumbler gears (operated by H5) are provided between the spindle and gear train along with a quadrant plate that enables a gear train of the correct ratio and direction to be introduced.
   在主轴和传动机构连同挂轮板之间由摆动换向齿轮(由 H5 控制)来选择正确的速

## Lesson 1　Lathe（metal）

比和螺距。

**语法补充：被动语态与各种时态的结合方式。**

一、本语法在注释七中的应用

　　注释七中"are provided"和"be introduced"语态都是被动语态，因为主语都是被动地接受某动作，且时态都是一般现在时，原因是这句话讲的是一种操作方法，而不是具体的一次事件。

二、对本语法的详细阐述

|  | 现在 | 过去 | 将来 | 过去将来 |
|---|---|---|---|---|
| 一般 | is/am/are done | was/were done | will/shall be done | would/should be done |
| 进行 | is/am/are being done | was/were being done | / | / |
| 完成 | have/has been done | had been done | will/shall have been done | would/should have been done |
| 完成进行 | / | / | / | / |

1. 一般现在时（am/is/are + done）

   e.g. English is spoken by lots of people in the world.

   　　世界上的许多人都说英语。

2. 一般过去时（was/were + done）

   e.g. The cup was broken by the boy.

   　　杯子被那个男孩打碎了。

3. 一般将来时与过去将来时（will/shall be + done; would/should be + done）

   e.g. A new road will be built next year.

   　　明年要修一条新马路。

   e.g. I thought thousands of people would be helped.

   　　我认为将有数千人得到帮助。

4. 现在进行时与过去进行时（am/is/are being + done; was/were being + done）

   e.g. The machine was being repaired at this time yesterday.

   　　昨天这时，机器正在被修理。

   e.g. The problem is being discussed now.

   　　问题正在被讨论。

5. 现在完成时（have/has been + done）

   e.g. Two hundred trees have been planted by now.

   　　到现在为止，已经种了二百棵树了。

6. 过去完成时(had been + done)

e. g. They said they had been invited to the party.

他们说已经被邀请参加晚会了。

7. 将来完成时(will/shall have been done)

e. g. The snow will have disappeared before the end of February.

积雪将于2月底之前融化。

8. 过去将来完成时(would/should have been done)

e. g. He said he would have finished his work the next day.

他说他第二天会完成工作。

## Check your understanding

Ⅰ. Give brief answers to the following questions.

1. At the least, what do mental lathes consist of?
2. What is the function of the bed?

Ⅱ. Match the items listed in the following two columns.

| | |
|---|---|
| toolroom lathe | 卡盘 |
| fine feed | 钻孔 |
| boring | 驱动杆 |
| quadrant plate | 刀架,刀座 |
| tumbler gear | 工具车床 |
| tool post | 挂轮板 |
| chuck | 微小进给 |
| driveshaft | 摆动换向齿轮 |
| pulley | 进给螺杆 |
| feedscrew | 皮带轮 |

## 车 床

金属车床或精工车床是对那些精加工相对较硬的材料的一类车床的通称。它们最初设计出来是用于加工金属材料,然而,随着塑料和其他材料的发展,车床的潜能逐渐地显现出来,用途越来越广泛,可以加工的材料也越来越多。在机械加工行话中,当金属车床的主要功能比较明确时,它们通常被简单称作车床或加上更专业的称呼,如工具车床或六角车床等等。这些刚性机床通过做直线运动的各种切削刀具(如车刀

# Lesson 1  Lathe（metal）

或钻头）来切削旋转运动的工件上的多余的材料。

## 结构

为适应各种加工用途,金属车床发生了很大的变化,可是,了解基本构造有助于了解它们的相似之处。这类机床的组成至少包括主轴箱,床身,拖板箱和尾架。这种改进的机床制造得很坚固,有着宽阔的支撑面（滑轨或导轨),使得它更稳定,制造精度更高。这将有助于确保其加工零件的制造公差和定位精度,满足要求。

## 主轴箱

主轴箱（H1）里有主轴（H4），变速手柄（H2、H3）和齿轮变速箱（H10）。在车削时会产生很大的切削力,切削力会使轻薄的箱体产生变形,并产生共振,振动会传到工件,影响工件的质量,所以主轴箱要有足够强度。

图 3.1-1  典型的车床

主轴通常是中空的,允许棒料扩展到工作区域,这降低了材料的制备和浪费。主轴在精密轴承的支撑下运转,其上还配备了附属的装夹仪器,比如卡盘和平面卡盘。主轴的末端也设计有一个锥孔,通常是莫氏锥度,以便锥柄或顶尖插入。老式车床的主轴由皮带轮直接带动,如果要获得更低的速度则要操纵大齿轮减速机构,后来车床的齿轮箱由专用的电动机带动。通过齿轮箱使这种动力头的速度选择进行得更加彻底。

图 3.1-2  主轴箱的基本组成

## 床身

结实的床身连接主轴箱并使拖板箱和尾座沿着床身上的导轨作平形于主轴方向的直线运动。供拖板箱和尾架移动的导轨是经过淬火和研磨的。拖板箱由齿轮齿条机构、准确螺距的丝杠或光杠带动。

### 光杠和丝杠

光杠(H8)是一根长长的轴,由齿轮控制它的转速,从而控制拖板箱的进退。这些齿轮位于床身下的拖板箱内。光杠和丝杠(H9)由(四分仪上的)变速齿轮或中间齿轮箱即快速进给齿轮箱(H6)或诺顿齿轮箱驱动。这些中间齿轮通过改变主轴和丝杠的速比来确定螺距和旋向来加工螺杆和蜗杆。在主轴和传动机构连同挂轮板之间由摆动换向齿轮(由 H5 控制)来选择正确的速比和螺距。这就在主轴转速和丝杠转速之间提供了一个常数关系。这个比率使得在加工螺杆时不需要模具的帮助。

丝杠将要被设计成既可加工英制螺纹又可加工公制螺纹,并且可以根据不同厂家的要求改变比率。为了使牙型之间的转换更加精确需要在车床上装一个 127 个齿的齿轮,如果车床上没有足够的地方装也可以用近似的方法。用 3 和 7 的倍数给一个比率 63:1 可以加工不太精确的牙型。这个变换比率通常设置在快速变速箱内。

### 拖板箱

简而言之,在操作者的操作下拖板箱带着刀架作纵向和横向运动。操作者可以通过手轮(5a)手动操纵拖板箱或者靠进给机构(5c)带动,当拖板箱由进给机构带动时,操作者会感觉操作起来很轻便,省力。为了能使用轻便和有助于重复加工,拖板箱上的手轮(2a,3b,5a)和导轨要经常校准。

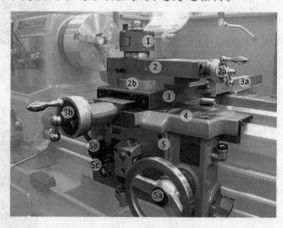

图 3.1-3 拖板箱的基本组成

### 横向导轨

横向导轨位于拖板箱的上方,它的光杠作与主轴轴线垂直的运动,这样可以完成端面车削。光杠和丝杠可以啮合起来(如前所述)使横向导轨自动运动,每次只能往一个方向运动,有一个自锁机构能关闭第二条传动链。

### 小刀架

小刀架(小拖板)是车床装刀架的一部分。它通过另一根光杠提供沿其轴线的小余量的运动。小刀架相对于拖板箱和导轨可以独自调整其位置。这在加工锥面,加工螺纹或想获得微小进给时就很有用了。

早在 15 世纪就有滑动刀架的记载了。Henry Maudslay 发明了滑动刀架的说法很

# Lesson 1  Lathe (metal)

流行,但是事实并非如此,他也从没声明这一点。从 James Nasmyth 那儿传出的一种说法是 Maudslay 发明了滑动刀架;1841 年,James Nasmyth 在他的《滑轨原理》的摘要中含糊不清地提到的,然后,其他作者误认为 Maudslay 是滑动刀架的发明者,并传播了这一错误。Maudslay 确实对这个理念的传播有过帮助。当他还是一个男孩的时候他很可能在兵工厂看过刀架的工作过程。在 1794 年,当他为 Joseph Bramah 工作时,他做了一个,然后当他有了自己的工厂后,他把滑动刀架广泛应用于自己做的车床上,并出售它们。再加上他训练的工程师团队,滑动刀架被人们熟知,并被其他的生产车床的厂家模仿,然后覆盖到整个英国的机械厂。一种实用的,万能的螺纹车床把光杠、变速齿轮和滑动刀架三者结合在一起是 Maudslay 最大的成就。

第一个有关全金属滑动刀架车床的完整记载中指出,法国人 Jacques de Vaucanson 早在 1751 年左右就发明了这种车床。在 Maudslay 发明并完善他的滑动刀架之前就已经记录在百科全书里很长时间了。好像 Maudslay 在他设计滑动刀架之初出现错误的时候并不知道 Vaucanson 已经发明了滑动刀架,因为这些错误并没有出现在 Vaucanson 的车床中。

## 刀架

刀具安装在刀架上,这种刀架可能沿用了美式灯笼造型,即传统的四条边的方形式样,或者是快速换位式的就像以前描述过的转塔刀架。快速换位刀架的优势在于它能够提供尽可能多的刀具来参与加工,每次只提供一把刀的灯笼式的刀架或者是每次提供三把或四把刀的方形式样的刀架是不能比拟的。可更换的刀架可以安装中心高一样的所有类型的刀具,即使刀架从机床上卸下来。

## 尾座

尾座是一种夹具,直接安装在主轴上,与主轴箱相对。尾座上的主轴(T5)是不旋转的,它只是在光杠和手轮(T1)的作用下作纵向的直线运动。主轴上有一个锥孔用来装钻头、中心钻或其他刀具。尾座可以在床身上通过锁紧螺母(T6)在要求的位置上定位。尾座架(T4)可以自动补偿相对于主轴轴线的偏移,这在加工小锥面的时候很有用。

图 3.1-4  尾座的基本组成

在这张图片上显示在手轮和主轴之间有一个减速齿轮箱(T2),这只有在大型的中心车床上才有,这种车床在钻大孔的时候就需要这种额外的杠杆作用。

## Lesson 2  Milling Machine

Fig. 3.2 – 1  **An example of a CNC vertical milling center**

*1*  • Example of a CNC vertical milling center

*2*  • A CAD designed part (top) and physical part (bottom) produced by CNC milling center.

*3*  • A milling machine is a machine tool used for the shaping of metal and other solid materials. Its basic form is that of a rotating cutter which rotates about the spindle axis (similar to a drill), and a table to which the workpiece is affixed. In contrast to drilling, where the drill is moved exclusively along its axis, the milling operation involves movement of the rotating cutter sideways as well as 'in and out'. The cutter and workpiece move relative to each other, generating a toolpath along which material is removed. The movement is precisely controlled, usually with slides and leadscrews or analogous technology. Often the movement is achieved by moving the table while the cutter rotates in one place, but regardless of how the parts of the machine slide, the result that matters is the relative motion between cutter and workpiece. Milling machines may be manually operated, mechanically automated, or digitally automated via computer numerical control (CNC).

*4*  Milling machines can perform a vast number of operations, some of them with quite complex toolpaths, such as slot cutting, planing, drilling, diesinking, rebating, routing, etc. Cutting fluid is often pumped to the cutting site to cool and lubricate the cut, and to sluice away the resulting swarf.

### Types of milling machines

*5*  There are many ways to classify milling machines, depending on which criteria are the

# Lesson 2  Milling Machine

focus:

| Criterion | Example of classification scheme | Comments |
|---|---|---|
| Control | Manual; Mechanically automated via cams; Digitally automated via NC/CNC | In the CNC era, a very basic (基本的) distinction is manual versus CNC. Among manual machines, a worthwhile distinction is non-DRO-equipped versus DRO-equipped |
| Control (specifically among CNC machines) | Number of axes (e. g., 3-axis, 4-axis, or more); Within this scheme, also: • Pallet-changing versus non-pallet-changing • Full-auto tool-changing versus semi-auto or manual tool-changing | |
| Spindle axis orientation | Vertical versus horizontal; Turret versus non-turret | Among vertical mills, "Bridgeport-style" is a whole class of mills inspired by the Bridgeport original |
| Purpose | General-purpose versus special-purpose or single-purpose | |
| Purpose | Toolroom machine versus production machine | Overlaps with above |
| Purpose | "Plain" versus "universal" | A distinction whose meaning evolved over decades as technology progressed, and overlaps with other purpose classifications above; more historical interest than current |
| Size | Micro, mini, benchtop, standing on floor, large, very large, gigantic | |
| Power source | Line-shaft-drive versus individual electric motor drive | Most line-shaft-drive machines, ubiquitous circa 1880-1930, have been scrapped by now |
| Power source | Hand-crank-power versus electric | Hand-cranked not used in industry but suitable for hobbyist micromills |

## 1. Comparing vertical with horizontal

6  In the vertical mill the spindle axis is vertically oriented. Milling cutters are held in the spindle and rotate on its axis. The spindle can generally be extended (or the table can be raised/lowered, giving the same effect), allowing plunge cuts and drilling. There are two subcategories of vertical mills: the bedmill and the turret mill. Turret mills, like the ubiquitous Bridgeport, are generally smaller than bedmills, and are considered by some to be more versatile. In a turret mill the spindle remains stationary during cutting operations

and the table is moved both perpendicular to and parallel to the spindle axis to accomplish cutting. In the bedmill, however, the table moves only perpendicular to the spindle's axis, while the spindle itself moves parallel to its own axis. Also of note is a lighter machine, called a mill-drill. It is quite popular with hobbyists, due to its small size and lower price. These are frequently of lower quality than other types of machines, however.

7  A horizontal mill has the same sort of x-y table, but the cutters are mounted on a horizontal arbor across the table. A majority of horizontal mills also feature a +15/-15 degree rotary table that allows milling at shallow angles. While endmills and the other types of tools available to a vertical mill may be used in a horizontal mill, their real advantage lies in arbor-mounted cutters, called side and face mills, which have a cross section rather like a circular saw, but are generally wider and smaller in diameter. Because the cutters have good support from the arbor, quite heavy cuts can be taken, enabling rapid material removal rates. These are used to mill grooves and slots. Plain mills are used to shape flat surfaces. Several cutters may be ganged together on the arbor to mill a complex shape of slots and planes. Special cutters can also cut grooves, bevels, radii, or indeed any section desired. These specialty cutters tend to be expensive. Simplex mills have one spindle, and duplex mills have two. It is also easier to cut gears on a horizontal mill.

Fig. 3.2 – 2  **A miniature hobbyist mill lainly showing the basic parts of a mill**

## 2. Other milling machine variants and terminology

8  • Box or column mills are very basic hobbyist bench-mounted milling machines that feature a head riding up and down on a column or box way.

9  • Turret or vertical ram mills are more commonly referred to as Bridgeport-type milling machines. The spindle can be aligned in many different positions for a very versatile, if somewhat less rigid machine.

10  • Knee mill or knee-and-column mill refers to any milling machine whose x-y table rides up and down the column on a vertically adjustable knee. This includes Bridgeports.

11  • C-Frame mills are larger, industrial production mills. They feature a knee and fixed spindle head that is only mobile (移动) vertically. They are typically much more

# Lesson 2　Milling Machine

powerful than a turret mill, featuring a separate hydraulic motor for integral hydraulic power feeds in all directions, and a twenty to fifty horsepower motor. Backlash eliminators are almost always standard equipment. They use large NMTB 40 or 50 tooling. The tables on C-frame mills are usually 18″ by 68″ or larger, to allow multiple parts to be machined at the same time.

12　● Planer-style mills are large mills built in the same configuration as planers except with a milling spindle instead of a planing head. This term is growing dated as planers themselves are largely(大量地,很多地) a thing of the past.

13　● Bed mill refers to any milling machine where the spindle is on a pendant(下垂物) that moves up and down to move the cutter into the work. These are generally more rigid than a knee mill.

14　● Ram type mill refers to a mill that has a swiveling cutting head mounted on a sliding ram. The spindle can be oriented either vertically or horizontally, or anywhere in between. Van Norman specialized in ram type mills through most of the 20th century, but since the advent of CNC machines ram type mills are no longer made.

15　● Jig borers are vertical mills that are built to bore holes, and very light slot or face milling. They are typically bed mills with a long spindle throw. The beds are more accurate, and the handwheels are graduated down to 0001″ for precise hole placement.

16　● Horizontal boring mills are large, accurate bed horizontal mills that incorporate many features from various machine tools. They are predominantly used to create large manufacturing jigs, or to modify large, high precision parts. They have a spindle stroke of several (usually between four and six) feet, and many are equipped with a tailstock to perform very long boring operations without losing accuracy as the bore increases in depth. A typical bed would have X and Y travel, and be between three and four feet square with a rotary table or a larger rectangle without said table. The pendant usually has between four and eight feet in vertical movement. Some mills have a large (30″ or more) integral(整体的) facing head. Right angle rotary tables and vertical milling attachments are available to further increase productivity.

17　● Floor mills have a row of rotary tables, and a horizontal pendant spindle mounted on a set of tracks(轨道) that runs parallel to the table row. These mills have predominantly been converted to CNC, but some can still be found (if one can even find a used machine available) under manual control. The spindle carriage moves to each individual table, performs the machining operations, and moves to the next table while the previous table is being set up for the next operation. Unlike any other kind of mill, floor mills have floor units that are entirely movable. A crane(起重机) will drop massive rotary tables, X-Y tables, and the like into position for machining, allowing the largest and most complex custom milling operations to take place.

*18* • A portical mills. It has the spindle mounted in a T structure.

## ❖ New Words and Phrases

| | |
|---|---|
| vertical ['və:tikəl] *adj.* | 垂直的 |
| exclusively [ik'sklu:sivli] *ad.* | 仅仅，专门，唯一地 |
| analogous [ə'næləgəs] *adj.* | 类似的 |
| rebate ['ri:beit, ri'beit] *n.* | 槽，榫头 |
| lubricate ['lu:brikeit] *v.* | 润滑，涂油 |
| routing ['ru:tiŋ] *n.* | 特形铣 |
| swarf [swɔ:f] *n.* | 木屑 |
| criteria [krai'tiəriə] *n.* | 标准 |
| pallet ['pælit] *n.* | 托盘 |
| overlap [,əuvə'læp] *n.* | 重叠，重复 |
| hobbyist ['hɔbiist] *n.* | 业余爱好者 |
| the turret mill | 转塔式铣床 |
| stationary ['steiʃ(ə)nəri] *adj.* | 不动的 |
| ubiquitous [ju:'bikwitəs] *a.* | 无所不在的，普遍存在的 |
| versatile ['və:sətail] *adj.* | 多方面的 |
| perpendicular [,pə:pən'dikjulə] *adj.* | 垂直的，直立的 |
| arbor ['ɑ:bə] *n.* | 藤架，凉亭 |
| mounted ['mauntid] *a.* | 安在马上的，裱好的 |
| aligned [ə'laind] *a.* | 排列的 |
| rigid ['ridʒid] *a.* | 刚性的 |
| mobile ['məubail] *a.* | 移动的 |
| appendant [ə'pendənt] *n.* | 下垂物 |
| integral ['intigrəl] *a.* | 整体的 |
| track [træk] *n.* | 轨道 |
| jig [dʒig] *n.* | 带锤子的钓钩 |
| borer ['bɔ:rə] *n.* | 钻孔器 |
| jig borer | 工模镗孔机 |
| crane [krein] *n.* | 起重机 |
| decade ['dekeid] *n.* | 十年 |

Lesson 2　Milling Machine

## ❖ Notes

1. In contrast to drilling, where the drill is moved exclusively along its axis, the milling operation involves movement of the rotating cutter sideways as well as 'in and out'.
钻削的运动(进给)始终是沿着轴线方向,铣削的运动包括旋转刀具的侧向和向上向下走刀。

2. Milling machines can perform a vast number of operations, some of them with quite complex toolpaths, such as slot cutting, planing, drilling, diesinking, rebating, routing, etc. Cutting fluid is often pumped to the cutting site to cool and lubricate the cut, and to sluice away the resulting swarf.
铣床可以完成很多种操作,其中有很多走刀路径都很复杂。像铣槽,铣平面,钻孔,模具铣削、铣榫头、特形铣等等。切削液往往是喷在加工作用点上起冷却和润滑作用的,并冲走切屑。

3. C-Frame mills are larger, industrial production mills. They feature a knee and fixed spindle head that is only mobile (移动) vertically. They are typically much more powerful than a turret mill, featuring a separate hydraulic motor for integral hydraulic power feeds in all directions, and a twenty to fifty horsepower motor. Backlash eliminators are almost always standard equipment. They use large NMTB 40 or 50 tooling.
C 型框架铣床比较大,是工业生产的铣床,它的特征是有一个膝型床身和一个固定的主轴,该主轴只能垂直移动。C 型框架铣床要比转塔铣床更有力。它的特征是在整体的液压系统的作用下,每个方向上都有独立的液压马达驱动进给,马达的功率在20~50马力。Backlash eliminators 几乎是标准的装备,它采用大型 NMTB40 或 50 机床。

**语法补充:非谓语动词中现在分词的用法**
一、本语法在注释三中的应用
　　本注释中"featuring"就是现在分词,由于"They are typically much more powerful than a turret mill"和"featuring a separate hydraulic motor for integral hydraulic power feeds in all directions"之间既没有句号隔开,又没有并列连词,且后者不是从句,因此两者中只能出现一个正常形态的谓语,即"are",若再想用动词,就要动用非谓语动词形式。此处由于语态为主动态,因此,使用-ing 形式的非谓语动词,即"featuring"。
二、对本语法的详细阐述
　　现在分词:表示正在进行的动作和主动语态(主动关系是与前面紧挨着的主语)。
e. g. The assistant serving her did not like the way she was dressed.

这名为她服务的店员不喜欢她的穿着。

在这句话中，serving 这一现在分词作为定语，修饰 the assistant，the assistant 与 serving 之间的关系为主动。

4. The tables on C-frame mills are usually 18″ by 68″ or larger, to allow multiple parts to be machined at the same time.

   C 型框架铣床上的工作台通常有 18″×68″或更大，它允许多个零件同时加工。

5. Bed mill refers to any milling machine where the spindle is on a pendant(下垂物) that moves up and down to move the cutter into the work. These are generally more rigid than a knee mill.

   床式铣床相对于其他铣床来说，主轴在一个下垂物上，可上下运动以切入工件，该类铣床一般比膝型铣床的刚性好。

## *Check your understanding*

Ⅰ. Give brief answers to the following questions.

1. What is the basic form of a milling machine?
2. What are the subcategories of vertical mills?

Ⅱ. Match the items listed in the following two columns.

| | |
|---|---|
| the turret mill | 工模镗孔机 |
| knee mill | 起重机 |
| crane | 圆柱形铣床 |
| floor mill | 转塔式铣床 |
| jig borer | 膝型铣床 |
| bed mill | 维修型铣床 |
| ram type mill | 盒型铣床 |
| column mill | 床式铣床 |
| box mill | 落地铣床 |

## 铣　　床

图中以一个立式数控铣削加工中心为例，顶部是由 CAD 设计的零件，底部是由数控加工中心制造出来的。

铣床是用来切削金属和其他一些固体材料的。它的基本组成有旋转刀具（绕着轴旋转,就像钻头)和用来固定住工件的工作台。铣削相比钻削,钻削的运动(进给)始终是沿着轴线方向,铣削的运动包括旋转刀具的侧向和向上向下走刀。刀具与工件间有相对运动,因此,走刀路径是沿着材料被切削的方向。运动被精确地控制,常常运用滑台、丝杆以及类似技术。通常是通过旋转的刀具不动,移动工作平台来实现的,但无论机床的各部件如何滑动,最后要保证的是刀具和工件间的相对运动。铣床可以通过手工操作,机械自动操作,或是计算机自动化数控操作。

图 3.2-1 立式数控铣削加工中心

铣床可以完成很多种操作,其中有很多走刀路径都很复杂。像铣槽,铣平面,钻孔,模具铣削、铣榫头、特形铣等等。切削液往往是喷在加工作用点上起冷却和润滑作用的,并冲走切屑。

**铣床的种类**

铣床的分类标准有多种,主要分类标准如下:

| 标　准 | 分类方案举例 | 注　释 |
|---|---|---|
| 控制 | 手工<br>通过 CAM 机械自动控制<br>通过 CNC/NC 数字化控制 | 在数字化的时代,一个最基本的的区别是手工对应数字控制。<br>在手工机器中,有无装备 DRO 装备是有明显区别的 |
| 通过 CNC 控制 | 轴数(例:3 轴 4 轴或者更多)<br>· 货盘转换和非货盘转换<br>· 全自动换刀和半自动或手动换刀 | |
| 轴的方向 | 立式与卧式;转塔式或非转塔式 | 在立式铣床中,Bridgeport 铣床是一个很有创造力的系列,最早由 Bridgeport 发明的 |

| 用途 | 普通型、特殊型、单一型 | |
|---|---|---|
| 目的 | 工作室机型与生产机型 | 同上 |
| 应用范围 | 简单的与通用的 | 数十年来，随着科技的进步，这一区别有不同意义，除去以上提到的共同点，主要是具有更多的历史意义 |
| 尺寸 | 微型，迷你型，台面型，独立放置型，大型，特大型，巨大型 | |
| 能源 | 总轴传动与独立电驱动 | 1880～1930年总轴传动被广泛采用，现已被废弃 |
| | 手动提供能量与电能 | 手工不会被工业采用，但业余爱好迷你铣床可以考虑使用 |

## 1. 立式与卧式的比较

立式铣床的主轴是竖直放置的。铣刀安装在主轴上并可以绕其旋转。轴可以伸缩（或者说工作平面可以升降，从而获得相同的效果），使得铣床可以进行深度切割和钻孔。立式铣床有两种主要的类型:床式和转塔式。转塔式铣床普遍采用的桥式，一般比床式铣床要小，而且功能更多。转塔式铣床的轴在切削过程中保持静止，台面会延垂直与平行于轴的方向完成切削。相比之下，床式铣削中，台面只会沿垂直于轴的方向运动，轴本身会沿平行于轴的方向运动。还有种值得注意的叫做铣钻的更轻的机床，由于造型小巧，往往成本低廉，在业余制造中经常被使用，但产品质量也较差。

图3.2-2 一个小型的业余加工机器 可以清楚的展示铣削机床的基础部件

卧式铣床拥有相同的X-Y坐标的工作台，但刀具安装在一个与工作台面垂直相交的水平柄轴上。大多数的卧式铣床工作台可以有+15至-15度的旋转，使得切削可以有少量的角度变化。一些可以进行垂直铣削的刀具也可以应用于卧式铣床，它们最大的优势就是拥有心轴安装的铣刀，如三面刃铣刀，它们的横截面像个圆锯，但宽度更大和直径更小。因为心轴的承载能力强，可以承受大切削量和快速走刀的切割。这些通常是用来铣槽的。平面铣削往往用来改变比较平整的表面，几把刀具可以组合起来在心轴上加工一个结构复杂的狭槽或平面。专用刀具可以切割沟槽斜面，弧形角，这些专业的刀具都是非常昂贵的。简单的铣床只有一个轴，双轴铣床的有两个轴。用卧式铣床加工齿轮相对更容易些。

## 2. 其他种类的铣床与术语

· 盒形与圆柱形铣床是最基本的业余爱好者使用的安装在板凳上的铣床,它的特征是铣削头沿圆柱或框架上下移动走刀就能进行加工。

· 转塔式铣床或立式维修式铣床,其主轴可排列在许多不同位置以保证它的通用性,但刚性不好。

· 膝型铣床或膝型导柱式铣床指任何一种有 X-Y 工作台,工作台能在沿垂直可调的膝型床身的导柱上下滑动的铣床。桥式铣床也是这样。

· C 型框架铣床比较大,是工业生产的铣床,它的特征是有一个膝型床身和一个固定的主轴,该主轴只能垂直移动。C 型框架铣床要比转塔铣床更有力。它的特征是在整体的液压系统的作用下,每个方向上都有独立的液压马达驱动进给,马达的功率在 20~50 马力。Backlash eliminators(齿轮侧隙消除机构)几乎是标准的装备,它采用大型 NMTB40 或 50 机床,C 型框架铣床上的工作台通常有 18″×68″或更大,它允许多个零件同时加工。

· 龙门刨铣床更大,它的特征是和龙门刨的外形相同,不同之处在于它用一个铣削头取代刨削头。从某种意义上来说,随着时间的推移刨削本身很多时候变成过去的事物。

· 床式铣床的主轴在一个下垂物上,可上下运动以切入工件,该类铣床一般比膝型铣床的刚性好。

· 维修型铣床是一种安装在滑台上的铣削头可旋转铣床,主轴可以水平也可以垂直,或在介于两者中间任一位置。Van Norman(冯.诺曼)在 20 世纪大部分时候专业生产这类铣床,但自从有了 CNC 机床以后,维修型铣床就不再生产了。

· 杰格镗床是一种立式铣床,主要用于钻孔、铣小槽或铣面。它是典型的床式铣床,有一个长的主轴行程。床身精度较高,在精确钻孔时,手轮每次向下进给精度可达.0001″。

· 卧式钻铣床较大,精密的床身和水平的铣削头使它具有各种铣床的特征,主要用来加工大型夹具或较大、精度高的零件。主轴有几种(通常 4~6 英尺),许多这种铣床配有尾架,用来加工比较深的孔,使得加工深孔时不致丧失精度。典型的床身有 X 和 Y 方向的导轨,大小在 3~4 平方英尺,一个可旋转的工作台;或有不回转的矩形工作台。下垂物(主轴箱)有 4 到 8 种在垂直方向上的进给速度。一些铣床有一个较大(30″或更大)整体面加工铣削头。可右旋的工作台、立铣头可进一步提高生产率。

· 落地铣床有一排旋转工作台,一根水平布置的主轴可在平行于一排工作台的导轨上运动。这类铣床大多有 CNC,但(在旧铣床里可以)找到手动控制的,主轴箱移动到每个工作台,完成铣削加工,然后再移动到下一个工作台,此时前面的工作台可以装卸工件,为后面的铣削做准备。和其他类型的铣床不同,落地铣床拥有可整体移动的落地单元,起重机可起吊一个大型的旋转工作台,X-Y 工作台,安装在指定位置上,用来完成更大更复杂的铣削加工。

・Portical 铣床有一个安装在 T 型结构上的主轴。

# Lesson 3  Types of Milling Cutter

## End mill

*1*  End mills (middle row in image) are those tools which have cutting teeth at one end, as well as on the sides. The words end mill are generally used to refer to flat bottomed cutters, but also include rounded cutters (referred to as ball nosed) and radiused cutters (referred to as bull nose, or torus). They are usually made from high speed steel(HSS) or carbide, and have one or more flutes. They are the most common tool used in a vertical mill.

Fig. 3.3 - 1   Slot, end mill, and ballnose cutters

## Slot drill

*2*  Slot drills (top row in image) are generally two (occasionally three or four) fluted cutters that are designed to drill straight down into the material. This is possible because there is at least one tooth at the centre of the end face. They are so named for their use in cutting keyway slots. The words slot drill is usually assumed to mean a two fluted, flat bottomed end mill if no other information is given. Two fluted end mills are usually slot drills, three fluted sometimes aren't, and four fluted usually aren't.

## Roughing end mill

*3*  Roughing end mills quickly remove large amounts of material. This kind of end mill utilizes a wavy tooth form cut on the periphery. These wavy teeth form many successive cutting edges producing many small chips, resulting in a relatively rough surface finish. During cutting, multiple teeth are in contact with the workpiece reducing chatter and vibration. Rapid stock removal with heavy milling cuts is sometimes called hogging. Roughing end mills are also sometimes known as ripping cutters.

## Ball nose cutter

*4*  Ball nose cutters (lower row in image) are similar to slot drills, but the end of the cutters are hemispherical. They are ideal for machining 3-dimensional contoured shapes in machining centres, for example in molds and dies. They are sometimes called ball mills in shop-floor slang, despite the fact that that term also has another meaning. They are also used to add a radius between perpendicular faces to reduce stress concentration.

## Slab mill

5   Slab mills are used either by themselves or in gang milling operations on manual horizontal or universal milling machines to machine large broad surfaces quickly. They have been superseded by the use of Carbide tipped face mills that are often used in vertical mills or machining centres.

Fig. 3.3 - 2  HSS slab mill

## Side-and-face cutter

6   The side-and-face cutter is designed with cutting teeth on its side as well as its circumference. They are made in varying diameters and widths depending on the application. The teeth on the side allow the cutter to make unbalanced cuts (cutting on one side only) without deflecting the cutter as would happen with a slitting saw or slot cutter (no side teeth).

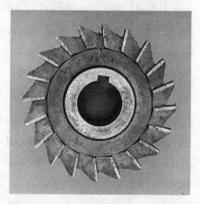

Fig. 3.3 - 3  Side and face cutter

## Involute gear cutter

7   The image shows a Number 4 cutter from an involute gear cutting set. There are 8 cutters (excluding the rare half sizes) that will cut gears from 12 teeth through to a rack (infinite diameter). The cutter shown has markings that show it is a

8   • 10 DP (diametrical pitch) cutter

9   • That it is No. 4 in the set

10  • that it cuts gears from 26 through to 34 teeth

11  • It will cut gears with teeth giving the gear a 14.5 degree pressure angle

Fig. 3.3 - 4  Involute gear cutter - No. 4

# Lesson 3  Types of Milling Cutter

## Hobbing cutter

12  These cutters are a type of form tools and are used in hobbing machines to generate gears. A cross section of the cutters tooth will generate the required shape on the workpiece, once set to the appropriate conditions (blank size). A hobbing machine is a specialised milling machine.

Fig. 3.3 – 5  **Hobbing cutter**

## Face mill (indexable carbide insert)

## Carbide tipped face mill

13  A face mill consists of a cutter body (with the appropriate machine taper) that is designed to hold multiple disposable carbide or ceramic tips or inserts, often golden in color. The tips are not designed to be resharpened and are selected from a range of types that may be determined by various criteria, some of which may be: tip shape, cutting action required, material being cut. When the tips are blunt, they may be removed, rotated (indexed) and replaced to present a fresh, sharp face to the workpiece, which increases the life of the tip and thus their economical cutting life.

## Fly cutter

14  A fly cutter is composed of a body into which one or two tool bits are inserted. As the entire unit rotates, the tool bits take broad, shallow facing cuts. Fly cutters are analogous to face mills in that their purpose is face milling and their individual cutters are replaceable. Face mills are more ideal in various respects (e.g. rigidity, indexability of inserts without disturbing effective cutter diameter or tool length depth-of-cut capability), but tend to be expensive, whereas fly cutters are very inexpensive.

Fig. 3.3 – 6  **Various sizes of woodruff key cutters and keys**

## Woodruff cutter

15  Woodruff cutters make the seat for woodruff keys. These keys retain pulleys on shafts and are shaped as shown in the image.

## Hollow mill

16  Hollow milling cutters, more often called simply hollow mills, are essentially "inside-out endmills". They are shaped like a piece of pipe (but with thicker walls), with their cutting edges on the inside surface. They are used on turret lathes and screw machines as an alternative to turning with a box tool, or on milling machines or drill presses to finish a cylindrical boss (such as a trunnion).

### ❖ New Words and Phrases

| | |
|---|---|
| rounded cutter | 圆周切削铣刀 |
| radius ['reidjəs] n. | 半径 |
| radiused cutter | 径向切削铣刀 |
| slot [slɔt] n. | 狭缝 |
| drill [dril] n. | 钻孔机,钻子 |
| slot drills | 键槽铣刀 |
| keyway ['kiːˌwei] n. | 键沟 |
| keyway slots | 键槽 |
| roughing end mill | 粗加工端铣刀 |
| wavy ['weivi] adj. | 有波浪的(锯齿状) |
| periphery [pə'rifəri] n. | 外围,圆周 |
| workpiece ['wəːkpiːs] n. | 工件 |
| ball nose cutters | 球头刀 |
| slab [slæb] n. | 平板 |
| HSS slab mill | 高速钢平面铣刀(圆柱螺旋刃铣刀) |
| side and face cutter | 三面刃铣刀 |
| involute ['invəluːt] n. | 渐伸线 |
| involute gear cutter - No. 4 | 渐开线齿轮铣刀(模数铣刀) |
| carbide ['kɑːbaid] n. | 碳化物 |
| tip [tip] v. | 装顶端 |
| carbide tipped face mill | 硬质合金刀片面铣刀 |
| resharpen [riː'ʃɑːp(ə)n] vt. | 再次磨尖 |
| fly cutter | 飞刀 |
| indexability ['indeksəbiliti] n. | 指望 |
| woodruff ['wudrʌf] n. | 半圆 |
| woodruff key cutter | 半圆键铣刀 |

## Lesson 3　Types of Milling Cutter

| | |
|---|---|
| hollow milling cutter | 内孔铣削刀具 |
| turret ['tʌrit] n. | 小塔 |
| turret lathes | 转塔车床 |
| screw [skru:] n. | 螺丝钉 |
| screw machine | 螺纹加工机床 |
| cylindrical boss | 圆柱形工件 |
| trunnion ['trʌnjən] n. 枢轴 | |

### ❖ Notes

1. A face mill consists of a cutter body (with the appropriate machine taper) that is designed to hold multiple disposable carbide or ceramic tips or inserts, often golden in color.

   该面铣刀含有刀体(有适当的机床锥度-莫氏锥度)，刀体用来固定多个可调节的硬质合金刀片或陶瓷刀片(通常是金黄色)及衬垫。

**语法补充：关系代词 that 的用法**

一、本语法在注释一中的应用

　　注释一中 that 是一个关系代词，引导 "that is designed to hold multiple disposable carbide or ceramic tips or inserts, often golden in color" 这个定语从句，修饰 "a cuter body" 这个先行词。

二、对本语法的详细阐述

　　关系代词 that 可以引导限制性定语从句，修饰代表人或事物的先行词，但不能用于引导非限制性定语从句。that 可以充当从句的主语、宾语、表语。

e. g. The bag that lies on the ground is hers.

　　地上的那个包是她的。(关系代词 that 指代 bag，在定语从句中充当主语)

e. g. The old man that I visited yesterday is my teacher.

　　我昨天拜访的那个老人是我的老师。(关系代词 that 指代 man，在定语从句中充当宾语)

e. g. He is no longer the star that he was.

　　他不再是过去的那位明星了。(关系代词 that 指代 the star，在定语从句中充当 was 的表语)

2. Face mills are more ideal in various respects (e. g., rigidity, indexability of inserts without disturbing effective cutter diameter or tool length offset, depth-of-cut capability), but tend to be expensive, whereas fly cutters are very inexpensive.

平面铣刀更能适用于不同的场合(例如:刚度,在不影响刀具有效直径以及长度、切削深度的能力的前提下),但是价格也是比较昂贵的,然而飞刀价格比较便宜些。

## *Check your understanding*

Ⅰ. Give brief answers to the following questions.
1. What does the end mill include?
2. What are the functions of the woodruff cutters?

Ⅱ. Match the items listed in the following two columns.

| | |
|---|---|
| rounded cutters | 螺纹加工机床 |
| screw machines | 径向切削铣刀 |
| radiused cutters | 键槽铣刀 |
| keyway slots | 圆周切削铣刀 |
| slot drills | 粗加工端铣刀 |
| roughing end mills | 半圆键铣刀 |
| carbide tipped face mill | 键槽 |
| cylindrical boss | 硬质合金刀片面铣刀 |
| woodruff key cutters | 圆柱形工件 |

# 铣刀的类型

### 立铣刀

立铣刀(图中间一排)是一端以及边上有切削齿的刀具。立铣刀这个称呼通常指平底铣刀,但也包括圆周切削铣刀(即球头刀)和径向铣刀(即牛鼻铣刀或圆环铣刀)。立铣刀通常由高速钢(HSS)或硬质合金制成,并有一个或多个排屑槽,是立式铣床上最常见的工具。

### 槽钻(键槽铣刀):

槽钻(图中第一排)一般有两个(有时 3 或 4 个)切削槽,可以直接钻入到材料中。这是可以做到的,因为至少有一个齿位于端面的中心。因为主要是用来铣削键槽故得名槽钻。如不做额外说明,槽钻通常是两个排屑槽的平底立铣刀,两个槽的立铣刀一般是槽钻,三

图 3.3-1 键槽铣刀、立铣刀、球头铣刀

个槽的立铣刀有时不是槽钻,四个槽的通常不是槽钻。

**粗铣立铣刀**

粗加工的立铣刀能迅速切除大量的材料。这种铣刀采用了锯齿状的齿,利用齿的边缘进行切削。这些锯齿状的齿形成许多连续的切削边界,能将材料切成很多的碎片,从而生成一个相对粗糙的表面质量。在切削过程中,有多个齿和工件接触从而减少了颤动。这种快速切成毛坯的粗铣有时被称为 hogging(扰乱)。粗铣立铣刀有时也被称为 ripping(撕)刀具。

**圆鼻刀**

圆鼻刀(图中下排)和槽钻很相似,但是刀具的尾部是半球体。在加工中心上,圆鼻刀是理想的加工三维轮廓形状的工具,例如模具。在工厂车间的俗语中,圆鼻刀有时被称为球刀,尽管球刀是另外一种刀具。圆鼻刀还可以在垂直面之间加工出圆角以减少应力集中。

**平面铣刀(螺旋刃圆柱铣刀)**

平面铣刀用于快速加工比较宽的平面,可以单独使用也可以在普通手控卧式铣床或万能铣床上进行组合铣削加工。平面铣刀已经被硬质合金镶尖的平面铣刀所替代,该铣刀常用于立式铣床和加工中心。

图 3.3-2　HSS 平面铣刀

**三面刃铣刀**

该铣刀的侧面和圆周上都有切削齿。它们根据不同的应用做出不同的直径和宽度。侧面的齿可以使刀具进行不对称铣削(只铣一边)而不用像开槽锯或切槽刀(无边齿)那样使刀具倾斜。

**渐开线圆柱齿轮铣刀//模数铣刀**

如图所示,该刀具是渐开线齿轮加工设备上的第 4 套铣刀。该设备共有 8 套铣刀(不包括罕见的缩小一半尺寸的铣刀)可以将齿轮从齿条(无限直径)铣削出 12 个齿。图示铣刀的标记有以下含义:

- 直径节距为 10
- 加工设备中的第四把刀

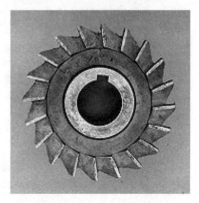

图 3.3-3　三面刃铣刀

- 加工齿数范围 26~34
- 可将齿轮铣削出14.5度压力角

### 滚刀

这些刀具是一种成形刀具,用于滚齿机加工齿轮。只要加工条件(毛坯尺寸)合适,该铣刀的刀齿横截面可以加工出所需形状的工件。滚齿机是一个专用铣床。

3.3-4 渐开线圆柱齿轮铣刀(第4套)

### 面铣刀(可转位硬质合金镶齿铣刀)

该面铣刀含有刀体(有适当的机床锥度-莫氏锥度),刀体用来固定多个可调节的硬质合金刀片或陶瓷刀片或衬垫(通常是金黄色)。刀尖选自一系列的标准形式,是不可以被磨砺的。刀尖所选的形式有以下决定因素:例如,刀尖形状,切削要求,材料。当刀尖钝了可以旋转至一个新的锋利的面进行加工。这样提高了刀尖的寿命以及经济性。

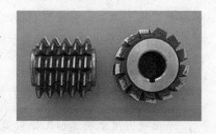

图 3.3-5 滚刀

### 飞刀

飞刀具由刀体以及装在上面的一个或两个刀头组成。当整套装置旋转时,刀头会进行宽而浅的平面铣削。飞刀和平面铣刀是相似的,它们共同的目的就是铣削平面,并且每一个刀片都可以替换。平面铣刀在诸多方面都更加理想(例如:刚度,在不影响刀具有效直径以及长度、切削深度的前提下的可转位能力),但是价格也是比较昂贵的,然而飞刀价格非常便宜。

### 半圆键铣刀

半圆键铣刀用来加工半圆键,这些键用于滑轮和传动轴之间的连接。形状如图所示。

### 内孔铣刀

内孔铣削刀具简称空心铣刀,实际是一种"外翻式立铣刀",他们形似一根管子(壁略厚),切削刃在内表面。这种刀具用于六角车床或螺丝车床中来代替组合刀具,或者用于铣床或钻床中完成圆柱

图 3.3-6 不同尺寸的半圆键铣刀和键

Lesson 3  Types of Milling Cutter

体的加工(如枢轴)。

**构词法之派生法**
派生法的构成方式(续)
(二) 后缀
3. 构成形容词的后缀
-ic，-ary，-cal，-ed，-tive，-able，-ible，-less
e. g.  macroscopic：[形容词] 宏观的
　　　revolutionary：[形容词] 革命的
　　　biological：[形容词] 生物学的
　　　based：[形容词] 以…为基础的
　　　unattractive：[形容词] 没有吸引力的
　　　repeatable：[形容词] 能够重复的
　　　reproducible：[形容词] 可再生的
　　　stainless：[形容词] 无污点的
4. 构成副词
-ly，-ward
e. g.  haphazardly：[副词] 随意地
　　　toward：[副词] 朝着(某个方向)
5. 构成数词
-teen（十几），-ty（几十），-th（构成序数词）。
e. g.  five 五→fifteen 十五→fifteenth 第十五→fifty 五十

# Unit 4

## Lesson 1  Engine Operating Principles

*1*　Most automobile engines are internal combustion, reciprocating 4-stroke gasoline engines, but other types have been used, including the diesel, the rotary (Wankel), the 2-stroke, and the stratified charge.

*2*　Reciprocating means "up and down" or "back and forth". It is the up and down action of a piston in the cylinder that produces power in a reciprocating engine. Almost all engines of this type are built upon a cylinder block, or engine block. The block is an iron or aluminum casting that contains engine cylinders and passages called water jackets for coolant circulation. The top of the block is covered with the cylinder head, which forms the combustion chamber. The bottom of the block is covered with an oil pan or oil sump.

Fig. 4.1 – 1  **Crankshaft connecting rod assembly**

*3*　Power is produced by the linear motion of a piston in a cylinder. However, this linear motion must be changed into rotary motion to turn the wheels of cars or trucks. The piston is attached to the top of a connecting rod by a pin, called a piston pin or wrist pin. The bottom of the connecting rod is attached to the crankshaft. The connecting rod transmits the up-and-down motion of the piston to the crankshaft, which changes it into rotary motion. The connecting rod is mounted on the crankshaft with large bearings called rod bearings. Similar bearings, called main bearings, are used to mount the crankshaft in the block.

# Lesson 1  Engine Operating Principles

Shown in Fig. 4.1 - 1.

*4*  The diameter of the cylinder is called the engine bore. Displacement and compression ratio are two frequently used engine specifications. Displacement indicates engine size. And compression ratio compares the total cylinder volume to compression chamber volume.

*5*  The term "stroke" is used to describe the movement of the piston within the cylinder, as well as the distance of piston travel. Depending on the type of engine the operating cycle may require either two or four strokes to complete. The 4-strole engine is also called Otto cycle engine, in honor of the German engineer, Dr. Nikolaus Otto, who first applied the principle in 1876. In the 4-stroke engine, four strikes of the piston in the cylinder are required to complete one full operating cycle. Each stroke is named after the action it performs: intake, compression, power, and exhaust in that order, shown in Fig. 4.1 - 2.

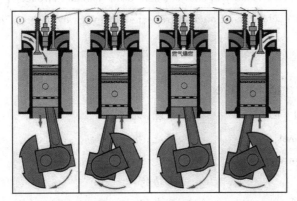

Fig. 4.1 - 2  Operating principle of 4 stroke engine

## 1. Intake stroke

*6*  As the piston moves down, the vaporized mixture of fuel and air enters the cylinder through the open intake valve. To obtain the maximum filling of the cylinder the intake valve opens about 10° before t. d. c. , giving 20° overlap. The inlet valve remains open until some 50° after b. d. c. to take advantage of incoming mixture.

## 2. Compression stroke

*7*  The piston turns up, the intake valve closes, the mixture is compressed within the combustion chamber, while the pressure rise to about 1MPa, depending on various factors including the compression ratio, throttle opening and engine speed. Near the top of the stroke the mixture is ignited by a spark which bridges the gap of the spark plug.

### 3. Power stroke

8  The expanding gases of combustion produces a rise in pressure of the gas to some 3.5MPa and the piston is forced down in the cylinder. The exhaust valve opens near the bottom of the stroke.

### 4. Exhaust stroke

9  The piston moves back up with the exhaust valve open some 50° before b. d. c. , allowing the pressure within the cylinder to fall and to reduce "back" pressure on the piston during the exhaust stroke, and burned gases are pushed out to prepare for the next intake stroke. The intake valve usually opens just before the exhaust stroke.

10  This 4-stroke cycle is continuously repeated in every cylinder as long as the engine remains running.

## ❖ New Words and Phrases

| | |
|---|---|
| principle ['prinsəpl] n. | 原理,机理 |
| combustion [kəm'bʌstʃən] n. | 燃烧,焚烧 |
| reciprocating [ri'siprəkeitiŋ] a. | 往复运动的 |
| rotary ['rəutəri] a. | 旋转的 |
| pan [pæn] n. | 平底锅 |
| oil pan | 油底壳 |
| pin [pin] n. | 大头针,针,拴 |
| piston pin | 活塞销 |
| bearing ['bɛəriŋ] n. | 轴承 |
| rod bearings | 连杆轴承 |
| displacement [dis'pleismənt] n. | 排气量 |
| compression [kəm'preʃ(ə)n] n. | 压缩 |
| ratio ['reiʃiəu] n. | 比,比率 |
| compression ratio | 压缩比 |
| volume ['vɔljuːm] n. | 体积,容积 |
| vaporized ['veipəraizd] a. | 蒸汽的,雾状的 |
| t. d. c. | 上止点 |
| throttle ['θrɔtl] n. | 节气门 |

# Lesson 1  Engine Operating Principles

## ❖ Notes

1. Most automobile engines are internal combustion, reciprocating 4-stroke gasoline engines.

   大多数汽车用发动机都是内部燃烧、往复运动四冲程汽油发动机。

2. The top of the block is covered with the cylinder head, which forms the combustion chamber.

   缸体的顶部盖着缸盖,它们一起组成了燃烧室。

3. To obtain the maximum filling of the cylinder, the intake valve opens about 10°before t. d. c., giving 20°overlap.

   为了尽可能的充满气缸,进气阀门在上止点前10°开启,这造成了20°的气门重叠角。

4. The piston turns up, the intake valve closes, the mixture is compressed within the combustion chamber, while the pressure rise to about 1MPa.

   活塞上行,进气门关闭,混合气被压缩在燃烧室里,此时压力上升至约1MPa。

### 语法补充:while 引导从句时的用法

一、本语法在注释四中的应用

　　本注释中,while 是一个关系副词,引导"while the pressure rise to about 1MPa"这个时间状语从句。

二、对本语法的详细阐述

1. while 引导时间状语从句

   e. g. The weather was fine while we were in Beijing.

   　　当我们在北京时,天气晴朗。

2. while 引导让步状语从句,表示"虽然,尽管"。

   e. g. While I admit that the problem is difficult, I don't think that they can't be solved.

   　　尽管我承认这个问题很难,但是我并不认为无法解决。

3. while 表示对比

   e. g. I earn only 120 dollars a week, while she earns 180 dollars.

   　　我一星期只赚120美元,她却赚180美元。

## *Check your understanding*

Ⅰ. Give brief answers to the following questions.

1. What are most automobile engines?
2. What is called the engine bore?
3. How many types of engines have been used? What are they?

4. How does the combustion chamber form?
5. How does the liner motion of the piston change into rotation motion?
6. What is the compression ratio?
7. Describe the four strokes.

Ⅱ. Match the items listed in the following two columns.

| | |
|---|---|
| throttle | 轴承 |
| oil pan | 燃烧 |
| combustion | 油底壳 |
| bearing | 压缩比 |
| displacement | 排气行程 |
| compression ratio | 活塞销 |
| piston pin | 节气门 |
| exhaust stroke | 排气量 |

# 发动机工作原理

大多数汽车用发动机都是内部燃烧、往复运动四冲程汽油发动机,但也有一些其他类型,如柴油机、转子式发动机(汪克尔)、二冲程发动机和复叠式进气发动机。

往复移动是指"上上下下"或"前前后后",在往复式发动机中正是气缸内活塞的这种上下运动产生了功率。几乎所有此类型的发动机都有赖于一个气缸体,或称机体。缸体由铸铁或铸铝制成,上面有气缸和水套,冷却水在水套内循环。缸体的顶部盖着缸盖,它们一起组成了燃烧室。缸体的底部装有油底壳,或称油盆。

气缸内活塞的直线运动产生功率。但是,这种直线运动必须转化为旋转运动才能驱动汽车车轮旋转。活塞依靠一个销与连杆头部相连,这个销称为活塞销或轴销。连杆下部与曲轴相连。连杆将活塞的上下运动传递给曲轴,曲轴将之转化为旋转运动。连杆安装在曲轴上,中间垫有大片轴承,称为连杆轴承。同样的轴承,装在曲轴和缸体之间的,称为主轴承。如图4.1-1所示。

气缸的直径称为发动机内径。两个常用的发动机参数是排气量和压缩比。排气量是指发动机容积,压缩比是指气缸总容积与压缩后燃烧室容积之比。

术语"冲程"用来表述气缸内活塞的运动,也用来表示活塞运行的路程。根据不同的发动机类型,完成一个工作循环可能需要两个或四个冲程。四冲程发动机也称为奥托循环发动机,这要感谢德国工程师,Nikolaus Otto博士,他在1876年首次应用了这个原理。在四冲程发动机里,气缸内活塞需要四个行程完成一整个工作原理。每个

Lesson 1　Engine Operating Principles

行程以它完成的工作来命名:进气、压缩、做功和排气,此过程如图4.1-2所示。

图4.1-1　曲轴连杆总成

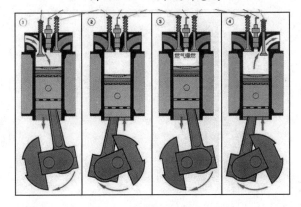

图4.1-2　四冲程发动机工作原理

### 1. 进气行程

当活塞向下运行时,燃油与空气的雾状混合物沿开启的进气阀门进入气缸。为了尽可能的充满气缸,进气阀门在上止点前10°开启,这造成了20°的气门重叠角。进气门一直开启到下止点后50°左右,这对利用进气混合是有利的。

### 2. 压缩行程

活塞上行,进气门关闭,混合气被压缩在燃烧室里,此时压力上升至约1MPa。此过程取决于多种因素,有压缩比,节气门开度和发动机转速。该行程接近顶点时,火花塞间隙处跳火,产生火花,点燃混合气。

### 3. 做功行程

燃烧产生的膨胀气体使气压上升至 3.5MPa 左右,气缸内活塞被压下,该行程接近终了时,排气阀门开启。

### 4. 排气行程

排气阀门在下止点前 50° 时开启,活塞随之上行,使缸内压力下降和减少了活塞上的"背"压,燃烧过的气体被推出,从而为下一个进气循环做准备。进气阀门通常在排气行程结束前就开启了。

发动机运转,则此四冲程循环在每个气缸中就持续重复进行。

# Lesson 2  Engine Construction

## Cylinder block

*1*  The cylinder block is cast in one piece. Usually, this is the largest and the most complicated single piece of metal in the automobile.

*2*  The cylinder block is a complicated casting made of gray iron (cast iron) or aluminum. It contains the cylinders and the water jackets that surround them. To make the cylinder block, a sand form called a mold is made. Then molten metal is poured into the mold. When the metal has cooled the sand mold is broken up and removed. This leaves the rough cylinder-block casting. The casting is then cleaned and machined to make the finished block. Fig. 4.2 – 1 shows the finished cylinder block.

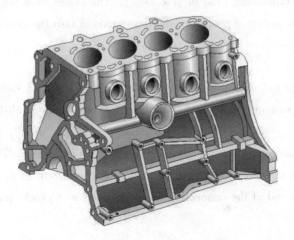

Fig. 4.2 – 1  **Cylinder block**

## Piston

*3*  The piston converts the potential energy of the fuel into the kinetic energy that turns the crankshaft. The piston is a cylindrical shaped hollow part that moves up and down inside the engine's cylinder. It has grooves around its perimeter near the top where the rings are placed. The piston fits snugly in the cylinder. The pistons are used to ensure a snug "air tight" fit. See Fig. 4.2 – 2.

*4*  The piston in your engine's cylinder is similar to your legs when you ride a bicycle.

Think of your legs as pistons; they go up and down on the pedals, providing power, Pedals are like the connecting rods; they are "attached" to your legs. The pedals are attached to the bicycle crank which is like the crank shaft, because it turns the wheels.

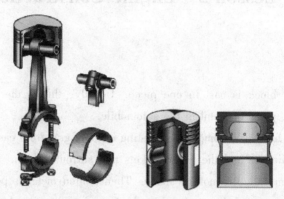

Fig. 4.2 - 2  **Piston**

5   To reverse this, the pistons (legs) are attached to the connecting rods (pedals) which are attached to the crankshaft (the bicycle rank). The power from the combustion in the cylinders powers the piston to push the connecting rods to turn the crankshaft.

## Connecting-rod

6   The connecting rod shown in Fig. 4.2 - 2 is made of forged high-strength steel. It transmits force and motion from the piston to the crank-pin on the crankshaft. A steel piston pin, or "wrist pin", connects the rod to the piston. The pin usually is pressed into the small end of the connecting-rod. Some rods have a lock bolt in the small end. As the piston moves up and down in the cylinder, the pin rocks back and forth in the hole, or bore, in the piston. The big end of the connecting rod is attached to a crank-pin by a rod bearing cap.

## Crankshaft

7   The crankshaft shown in Fig. 4.2 - 3 is the main rotating member, or shaft, in the engine. It has crank-pins, to which the connecting rods from the pistons are attached. During the power strokes, the connecting rods force the crank-pins and therefore the crankshaft to rotate. The reciprocating motion of the pistons is changed to rotary motion as the crankshaft spins. This rotary motion is

Fig. 4.2 - 3  **Crankshaft**

# Lesson 2  Engine Construction

transmitted through the power train to the car wheels.

## ❖ New Words and Phrases

| | |
|---|---|
| automobile [ˈɔːtəməubiːl] n. | 汽车,车辆 |
| gray iron | 灰铸铁 |
| water jacket | 水套 |
| rough [rʌf] a. | 粗糙的 |
| machine [məˈʃiːn] v. | 机加工,加工 |
| kinetic [kaiˈnetik] adj. | 运动的 |
| kinetic energy | 动能 |
| hollow [ˈhɔləu] a. | 中空的,空心的 |
| bearing cap | 轴承盖 |
| reciprocating [riˈsiprəkeitiŋ] n. | 往复 |
| crankshaft [ˈkræŋkʃɑːft] n. | 机轴 |

## ❖ Notes

1. The casting is then cleaned and machined to make the finished block.
   清洗和机加工后,就得到成品机体。
2. It has grooves around its perimeter near the top where the rings are placed.
   活塞顶部周边有一些凹槽,安装活塞环。
3. To reverse this, the pistons (legs) are attached to the connecting rods (pedals) which are attached to the crankshaft (the bicycle rank).
   反过来说,活塞(腿)与连杆(脚蹬子)相连,连杆(脚蹬子)又与曲轴(自行车曲柄)相连。
4. During the power strokes, the connecting rods force the crank-pins and therefore the crankshaft to rotate.
   在做功冲程,连杆驱动曲柄销,这样,曲轴就转起来了。

**语法补充:谓语的省略**
一、本语法在注释四中的应用
　　本注释中,"the crankshaft"后面省略了谓语"is forced",省略的原因是前文已经出现了相同的谓语,此处省略不会引起歧义,且使句子简洁。
二、对本语法的详细阐述
　　省略谓语的原因:后一句的谓语和前一句的谓语是一样的。

e. g. Only one of us was injured, and he just slightly.

我们当中只有一人受了伤,而且只是轻伤。

分析:he 后省去谓语 was injured。

e. g. We went through the tests on a Monday. Jenny had hers during the day, and I mine after work.

有一个星期一,我们进行了检查。詹尼在白天,我是在下班之后。

分析:I 之后省去谓语 had。

e. g. I'll be round as quick as I can. 我将尽快赶来。

分析:can 之后省去主要动词 be。

e. g. I pitied her sincerely, as I would a child of my own.

我真心地爱怜她,就像爱怜我自己的孩子一样。

分析:would 之后省去主要动词 pity。

## *Check your understanding*

Ⅰ. Give brief answers to the following questions.

1. What is the largest and the most complicated single piece of metal in the automobile?
2. What is the function of piston?
3. What is a cylinder block made of?
4. What is the function of the connecting rod?
5. What is the function of the crankshaft?

Ⅱ. Match the items listed in the following two columns.

| | |
|---|---|
| bearing cap | 灰铸铁 |
| rotary motion | 汽车、车辆 |
| automobile | 水套 |
| finished block | 动能 |
| kinetic energy | 成品机体 |
| water jackets | 轴承盖 |
| gray iron | 机轴 |
| crankshaft | 旋转运动 |

# 发动机构造

## 机体

气缸是整体浇铸。通常,它是汽车中最大最复杂的独立金属件。

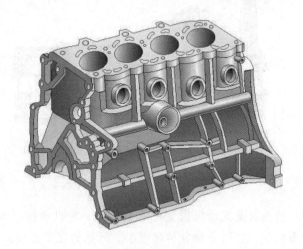

图 4.2 - 1 机体

机体是复杂的灰铸铁或铸铝件。内有气缸,气缸周围有水套。制造机体需要一套砂模,我们称之为模具。将金属溶液倒入模子,当金属冷却后,将砂模打碎除去,就得到了粗糙的气缸铸件。清洗和机加工后,就得到成品机体。如图 4.2 - 1 所示的缸体。

## 活塞

活塞将燃油的潜在能量转化为动能来驱动曲轴。活塞是圆柱形的空心零件,它在气缸内上下移动。活塞顶部周边有一些凹槽,安装活塞环。活塞在缸内紧密配合,用以确保气密性。见图 4.2 - 2。

发动机气缸内的活塞与你骑自行车时的腿很相象。把你的腿想象成活塞,腿在脚蹬子上上下运动,产生动力。脚蹬子就像连杆,它们与你的腿"相连",脚蹬子与自行车曲柄相连,曲柄可以转动车轮,就如同发动机曲轴一样。

反过来说,活塞(腿)与连杆(脚蹬子)相连,连杆(脚蹬子)又与曲轴(自行车曲柄)相连。缸内燃烧产生的功推动活塞,活塞推动连杆来转动曲轴。

## 连杆

图 4.2 - 2 所示的连杆由高强度的锻钢制成。它将来自活塞的力和运动传给曲轴

上的曲柄销。一根钢制的活塞销,或称销轴,连接连杆与活塞。通常将活塞销压入连杆小头。一些连杆的小头处有锁销。当活塞在缸内上下运动时,活塞销在活塞孔内前后晃动。连杆大头用一个连杆轴承盖与曲柄销相连。

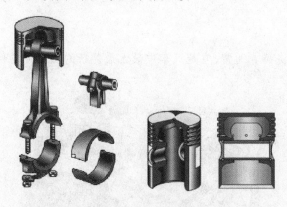

图 4.2-2 活塞

### 曲轴

图 4.2-3 所示的曲轴是发动机内重要的旋转件。它以曲柄销与连杆相连,在做功冲程,连杆驱动曲柄销,这样,曲轴就转起来了。活塞的往复直线运动就转化为旋转运动。这种旋转运动通过传动系统传递给车轮。

图 4.2-3 曲轴

### 构词法之转化法

一、定义:转化法指把一种词性用作另一种词性而词形不变。
二、具体构成方式
(一)动词转化为名词
process([动词] 加工)→process([名词] 加工工艺)
e. g. Plaster casting is an inexpensive alternative to other molding processes.

石膏铸造是一种替代其他铸造工艺的廉价铸造工艺。

(二)形容词转化为名词

additive([形容词]附加的)→additive([名词]添加剂)

e. g. Plastics are composed of polymer molecules and various additives.

　　塑料的组成成分是高聚物分子和各种添加剂。

(三)过去分词转化为名词

given([过去分词]被给予的)→ given([名词]理所当然的事情)

e. g. Deploying robotics in many industries is so routine that it is taken as a given.

　　在许多行业使用机器人是一件司空见惯、众所周知的事情。

# Unit 5

## Lesson 1  Forming of Sheet Metals

*1*  Metalworking is the process of working with metals to create individual parts, assemblies, or large scale structures. The term covers a wide range of work from large ships, bridges and oil refineries to delicate jewellery. It therefore includes a correspondingly wide range of skills and the use of many different types of metalworking processes and their related tools.

*2*  Metalworking generally is divided into the following categories, forming, cutting, and, joining. Each of these categories contain various processes.

*3*  Forming processes modify the shape of the object being formed by deforming the object, that is, without removing any material. It includes a collection of processes wherein the metal is rearranged into a specified geometry (shape) by:

*4*  • heating until molten, poured into a mold, and cooled, such as casting.

*5*  • heating until the metal becomes plastically deformable by application of mechanical force, such as forging.

*6*  • by the simple application of mechanical force, such as stamping.

*7*  Sheet metal is generally characterized by a high ratio of surface area to thickness. It is one of the fundamental forms used in metalworking, and can be cut and bent into a variety of different shapes. Countless everyday objects are constructed of the material. Thicknesses can vary significantly, although extremely thin thicknesses are considered foil or leaf, and pieces thicker than 6 mm (0.25 in) are considered plate. There are many different metals that can be made into sheet metal, such as: Aluminum, brass, copper, steel, tin, nickel and titanium. For decorative uses, important sheet metals include silver, gold, and platinum.

*8*  Metal stamping is a process employed in manufacturing metal parts with a specific design, by which sheets or strips of material are punched using a machine press or stamping press. Sheet metal can be molded into different pre-determined shapes. This could be a single stage operation where every stroke of the press produces the desired form on the sheet metal part, or could occur through a series of stages.

*9*  Sheet metal stamping dies are used to produce high precision metal components which are identical in shape and size. The dimensional accuracy and stability which you can achieve using precision metal stamping dies are very high and thus metal stamping dies are integral part of any manufacturing industry. You can see metal stamping components everywhere. From the electrical switches at your home to the computer you are surfing, from cars to aircrafts, everything needs precision metal stamping parts.

10    The industry of metal stamping has grown rapidly and, where possible, has replaced other metal forming processes like die casting, forging and machining for several reasons. The primary reason is cost effectiveness in deploying the stamping process. The dies used in metal forging and casting are more expensive than those used in metal stamping. The cost of the secondary processes, like cleaning and plating are greatly reduced.

11    A lot of basic and exotic metals can be used for stamping applications because of their malleable and ductile properties. The metal should not be very hard and ideally should have a low coefficient of flow. Some typical metals include:

12    • Ferrous metals: stainless steel stampings, and other iron-based metals

13    • Non-ferrous metals: brass stampings, bronze stampings, zinc stampings, and other metals that are not iron-based

14    • Exotic metals: beryllium copper, beryllium nickel, niobium, tantalum, and titanium stampings

15    • Precious metals: gold, silver, platinum, which are often used for decorative stampings

## ❖ New Words and Phrases

| | |
|---|---|
| forming ['fɔːmiŋ] n. | 成形,成型 |
| forging ['fɔːdʒiŋ] n. | 锻件,锻造(法) |
| stamping ['stæmpiŋ] n. | 冲压件(模锻,冲击制品) |
| sheet metal | 金属片,金属板 |
| foil [fɔil] n. | 箔,金属箔,薄金属片 |
| strip [strip] n. | 长条,条状 |
| punch [pʌntʃ] n. | 冲压机,冲头,冲孔,凸模冲头 vt.(用冲床)冲,冲孔 |
| a single stage | 单工位 |
| metal stamping die | 金属冲压模具 |
| plate [pleit] v. | 镀(覆以金属板,熨平) |
| plating ['pleitiŋ] n. | 电镀,镀敷 |
| malleable ['mæliəbl] a. | 可塑的,易改变的 |
| ductile ['dʌktail] adj. | 可锻的,有延展性的,韧性的 |

ferrous [ˈferəs] *a.*   含铁的

## ❖ Notes

1. Forming processes modify the shape of the object being formed by deforming the object, that is, without removing any material.
   成形过程是通过使物体变形,即采用不去除任何材料的方法,改变物体的形状。
2. Metal stamping is a process employed in manufacturing metal parts with a specific design, by which sheets or strips of material are punched using a machine press or stamping press.
   金属冲压成形是金属板料或条料在冲床或压力机的冲头压力作用下,生产具有特定设计要求的金属零件的一种成形方法。
3. This could be a single stage operation where every stroke of the press produce the desired form on the sheet metal part, or could occur through a series of stages.
   它通过压力机的每一次行程在金属板料零件上加工出所需的形状,这可以在一个单工位的操作中或通过一系列多工位的操作中完成。
4. The industry of metal stamping has grown rapidly and, where possible, has replaced other metal forming processes like die casting, forging and machining for several reasons.
   金属冲压工业已经加快了飞速发展的步伐,只要具备使用的可能性,冲压成形已拥有很多理由替代其他的金属成形过程,如铸造、锻造和机械加工等。

**语法补充:现在完成时的用法**

一、本语法在注释四中的应用
  本注释中"has grown"和"has replaced"都用了现在完成时,表示两个动作都是过去就已经发生,且对现在还有影响。
二、对本语法的详细阐述
1. 表示过去发生的动作,现在已经结束,但对现在有影响。
e. g. He has turned off the light.
  (灯已关)他已经把灯关了。
2. 表示过去发生的状态,现在还在持续。如注释四中"has grown"的用法。
e. g. I haven't seen him for a long time.
  我已经很长时间没有看见他了。

## *Check your understanding*

Ⅰ. Give brief answers to the following questions.

# Lesson 1  Forming of Sheet Metals

1. What is metalworking?
2. What is metalworking generally divided into?

Ⅱ. Match the items listed in the following two columns.

| | |
|---|---|
| stamping | 金属冲压模具 |
| jewellery | 金属片 |
| sheet metal | 冲压件 |
| plating | 稳定性 |
| metal stamping dies | 准确性 |
| exotic metals | 电镀 |
| accuracy | 特殊金属 |
| stability | 珠宝 |

## 金属板料的成型加工

金属加工是对金属材料进行加工,从而生产出独立的零件、装配体或者是规模很大的结构件的过程。金属加工形成的产品范围很广,从巨型轮船、桥梁、炼油设备到精致的珠宝。因此,它包括涉及相关领域很广的工艺技术,许多各不相同的金属加工方法以及与其相关的工具。

金属加工从总体上分成以下几类:金属成形,切割和连结。每一类都包括各种各样的加工过程。

成形过程是通过使物体变形,即不去除任何材料的方法,改变物体的形状。它包括一系列使金属转变成为特定几何形状的过程,如:

- 加热金属直至熔化,将其浇注到模具中,然后进行冷却,例如铸造成型;
- 加热金属至可塑性变形的状态,借助于机械作用力对金属进行成形,例如锻造成形;
- 仅仅使用机械作用力对金属进行成形,例如冲压;

金属板料的一般特征是表面积和厚度之间的比率很高。它是金属加工中工件的基本形式之一,能够被切割和弯曲成各种不同的形状。利用金属板料制成的各种日常用品不计其数。板料厚度可变化的范围很大,有的厚度极薄,如箔片一般。当其厚度超过6mm(或是0.25inch)时,我们才称之为金属板材。许多种类的金属可以制作成金属板料,如铝、黄铜、青铜、钢、锡、镍和钛等。用于装饰品的重要金属板料有银、金和铂。

金属冲压成形是金属板料或条料在冲床或压力机的冲头压力作用下,生产具有特定设计要求的金属零件的一种成形方法。金属板料在模具中能成型出各种各样的预先设定的形状。它通过压力机的每一次行程在金属板料零件上加工出所需的形状,这可以在一个单工位的操作中或通过一系列多工位的操作中完成。

金属板料冲压模具用来生产高精度的,在形状和尺寸上要求一致的批量金属零件。使用精密的金属冲压模具,获得的零件的尺寸准确性和稳定性会非常高。因此金属冲压模具是任何制造业中的重要组成部分。你可以在任何地方看见金属冲压零部件,从家中的电器开关到正在上网冲浪的计算机,从汽车到飞机,每一件物品上都需要高精度的金属冲压制件。

金属冲压工业飞速发展,只要条件许可,冲压成形即可代替其他的金属成形过程,如铸造、锻造和机械加工等。这么做有几点理由,最主要的原因是应用冲压成形成本低廉。金属锻造和铸造成形中使用的模具成本要比冲压成形的模具成本昂贵的多。辅助加工过程如清理和电镀等的成本也会大大地降低。

很多普通金属和特殊金属具有延展性和韧性,都能进行冲压加工。这类金属必须硬度不高而且原则上要求流动系数也较低。一些典型的金属包括:
- 含铁金属——不锈钢和其他铁基金属冲压件
- 非铁金属——黄铜、青铜、锌和其他非铁金属
- 特殊金属——铍铜、铍镍、铌、钽、钛
- 贵金属——金、银、铂等常用于装饰冲压物品

# Lesson 2   Metal Stamping Process and Die Design

1   Several metal stamping techniques are extensively used in industries and engineering applications. Blanking, piercing, bending, deep drawing, coining, embossing and cold extrusion are some of the examples of metal stamping techniques. Some of these are discussed below:

2   Bending: The bending operation is the act of bending blanks at a predetermined angle. An example would be an "L" bracket which is a straight piece of metal bent at a 90° angle. The main difference between a forming operation and a bending operation is the bending operation creates a straight line bend (such as a corner in a box) as where a form operation may create a curved bend.

3   Usually bending has to overcome both tensile stresses as well as compressive stresses. When bending is done, the residual stresses make it spring back towards its original position, so we have to overbend the sheet metal in order to overcome the residual stresses.

4   Blanking: A blanking die produces a flat piece of material by cutting the desired shape in one operation. The finish part is referred to as a blank. Generally a blanking die may only cut the outside contour of a part, often used for parts with no internal features.

5   Cut off: Cut off dies are used to cut off excess material from a finished end of a part or to cut off a predetermined length of material strip for additional operations.

6   Drawing: The drawing operation is very similar to the forming operation except that the drawing operation undergoes severe plastic deformation and the material of the part extends around the sides. A metal cup with a detailed feature at the bottom is an example of the difference between formed and drawn. The bottom of the cup was formed while the sides were drawn.

7   Piercing: The piercing operation is used to pierce holes in stampings.

8   A die set is the fundamental portion of every die. It consists of a lower shoe, or a die shoe, and an upper shoe, both machined to be parallel within a few thousandths of an inch. The upper die shoe is sometimes provided with a shank, by which the whole tool is clamped to the ram of the press. Because of their much greater weight, large dies are not mounted this way. They are secured to the ram by clamps or bolts. However, sometimes even large die sets may contain the shank, which in such a case is used for centering of the tool in the

press.

9   Figures 5.2-1 and 5.2-2 show the basic components of a compound and a progressive die. Both die shoes, upper and lower, are aligned via guide pins or guide posts. These provide for a precise alignment of the two halves during the die operation. The guide pins are made of ground, carburized, and hardened-tool steel, and they are firmly embedded in the lower shoe. The upper shoe is equipped with bushings into which these pins slip-fit. The die block, containing all die buttons, nests, and some spring pads, is firmly attached to the lower die shoe. It is made of tool steel, hardened after machining.

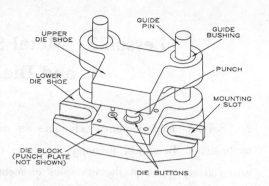

Fig. 5.2-1  **Basic components of a compound die**

10   The die block is usually a block of steel, either solid or sectioned, into which the openings are machined. The openings must match the outside shapes and outside diameters of the die bushings; they must be precise and exact, since the die bushings are press-fitted into them.

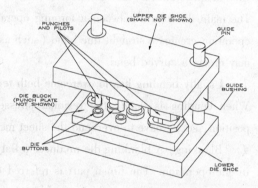

Fig. 5.2-2  **Basic components of a progressive die**

11   The punch plate is mounted to the upper shoe in much the same manner as the die block. Again, it is made of a hardened-tool steel, and it may consist of a single piece of steel, or be sectioned. It holds all punches, pilots, spring pads, and other components of the die. Their sizes and shapes conform to tooling they must contain minus the tolerance amount for press fit.

12   Both the die block and the punch plate are often separated from the die shoe by back-up plates, whose function is to prevent the punches and dies from becoming embedded in the softer die shoe. The sheet-metal strip is fed over the die block's upper surface, and it is usually secured between guide rails. There are two types of gauges: side gauges, for guiding the sheet through the die, and end gauges, which provide for the positioning of stock under the first piercing punch or blanking punch at the beginning of each strip. The strip is covered up, either whole or its portions, by the stripper, which provides for

stripping of the pierced material off the punch. The stripper is usually made from cold-rolled steel, and its openings are clearance openings for the shapes of punches.

13   The stationary stripper is mounted to the upper surface of the die block fastened with the same screws and dowels. The spring-loaded stripper is held in an offset location by the force of springs, and in such a case it is attached to the punch plate. With reverse punching, where the punch is mounted in the die block and the die is up in the punch plate, the stripping arrangement is reversed.

14   The cross-section of a typical die set is shown in Fig. 5.2 – 3. Here the knock out pins are going through the head of the punch, their stripping pressure being provided by a spring. The pins force the pressure pad or stripping insert out against the material, so that the blank is held down when the punch moves upward. Their pressure increases with the descent of the die. The die contains a similar set of pins, here called push pins. These lift up the cup off the die face after forming. The stripper is stationary, and it prevents the remainder of the strip from moving up on opening of the die, along with the movement of forming/blanking punch. This punch cuts the blank out of the strip with its outer diameter, forming it afterward with its face area and inner diameter's edge, finally bottoming on a forming support.

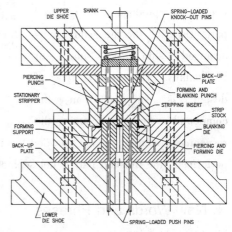

Fig. 5.2 – 3   **Cross-section of a typical die set**

15   There are considerable differences in the way dies are built to function. In some, the metal strip is fed through the die, which produces the desired part in stages. Another die makes a complete part with a single hit of a single station. According to their construction and function, all dies can be separated into the following four groups: compound dies, progressive dies, steel-rule dies, others dies.

16   Compound dies combine two or more operations at one station. A compound die performs only cutting operations (usually blanking and piercing) which are completed during a single press stroke. A characteristic of compound dies is the inverted position of the blanking die and blanking punch which also functions as the piercing die. As shown in Figure 5.2 – 4, the blanking die is fastened to the upper shoe and the blanking punch having a tapered hole in it and in the lower shoe for slug disposal is mounted on the lower shoe.

17  Progressive dies (shown earlier in Fig. 5.2-2) are a mixture of various single dies operating as different stations and grouped into the same die shoe. These stations are positioned to follow a sequence of operations needed to produce the required part. Usually, the die sequence is arranged side by side, or horizontally. This general type of design is simpler than the compound dies, because the respective operation are not crowded together. Regardless of the number of operations to be performed, the finished part is not separated from the strip until the last operation.

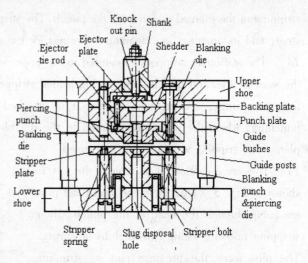

Fig. 5.2-4  A compound die

## ❖ New Words and Phrases

| | | |
|---|---|---|
| blanking [ˈblæŋkiŋ] n. | | 空白,下料,落料 |
| piercing [ˈpiəsiŋ] n. | | 冲孔加工 |
| deep drawing | | 深拉伸 |
| coining [ˈkɔiniŋ] n. | | 精压(立体挤压,压花,压印加工) |
| embossing [imˈbɔsiŋ] n. | | 浮雕(压纹,压花,模压加工) |
| extrusion [eksˈtruːʒən] n. | | 挤出 |
| cold extrusion | | 冷挤压 |
| tensile stress | | 拉应力 |
| compressive stress | | 压应力 |
| residual stress | | 残余应力 |
| die set | | 模架 |
| lower shoe | | 下模座 |
| upper shoe | | 上模座 |
| shank [ʃæŋk] n. | | 模柄 |
| clamp [klæmp] n. | | 夹具、锁紧 vt. 夹紧,固定 |
| die shoe | | 模座 |
| guide pin | | 导柱 |
| grind [graind] vt. | | 研磨 |

# Lesson 2  Metal Stamping Process and Die Design

| | |
|---|---|
| carburize [ˈkɑːbjuraiz] vt. | 使渗碳 |
| die block | 凹模固定板 |
| die button | 冲模凹模 |
| die bushing | 模套 |
| punch plate | 凸模固定板 |
| hardened-tool steel | 淬硬工具钢 |
| tolerance [ˈtɔlərəns] n. | 公差 |
| back-up plate | 支承垫板 |
| side gauge | 侧压板 |
| stripper [ˈstripə] n. | 卸料板 |
| cold-rolled steel | 冷轧钢 |
| stationary stripper | 固定式卸料板 |
| screw [skruː] n. | 螺丝钉 |
| dowel [ˈdauəl] n. | 销 |
| spring-loaded stripper | 弹压式卸料板 |
| reverse punching | 倒装冲裁模 |
| cross-section | 剖面,剖视图 |
| knock out pin | 打料推杆 |

## ❖ Notes

1. When bending is done, the residual stresses make it spring back towards its original position, so we have to overbend the sheet metal in order to overcome the residual stresses.

   当弯曲完成时,残余应力会使板料回弹至初始位置,因此须将金属板料加大弯曲变形程度以克服残余应力的影响。

## 语法补充:表示目的的英语表达法

一、本语法在注释一中的应用

　　本注释运用了 in order to do sth(为了做某事)这种表示目的的搭配,说明了"overbend the sheet metal"的目的是什么。

二、对本语法的详细阐述

(一)使用表示目的的短语如 in order to, so as to, to 等。

e.g. He is working hard in order to/so as to/to pass the examination.

　　他为了通过考试正在努力学习。

(二)使用目的状语从句

1. 用 in order that 引导：in order that 的意思是"为了"。

   e.g. He is working hard in order that he can pass the examination.

   为了考试及格,他正在努力学习。

2. 用 so that 引导：so that 此时的意思是"以便"。

   e.g. She burned the letters so that her husband would never read them.

   她把信都烧了,这样一来她丈夫就永远看不到了。

3. 用 in case 引导：in case 此时的意思是"以防""以备"。

   e.g. I always keep a bottle of wine in case friends call round.

   我平时总存着一瓶酒以备朋友来时喝。


2. The guide pins are made of ground, carburized, and hardened-tool steel, and they are firmly embedded in the lower shoe. The upper shoe is equipped with bushings into which these pins slip-fit.

   导柱选用的是经过研磨、渗碳、淬硬后的钢材,它们与下模座呈过盈配合。上模座配置了导套,导柱与导套呈间隙滑动配合。

3. The openings must match the outside shapes and outside diameters of the die bushings; they must be precise and exact, since the die bushings are press-fitted into them.

   这些孔必须和模套的外部形状和直径相匹配:它们的尺寸必须精确,才能保证模套压入后能配合完好。

4. The strip is covered up, either whole or its portions, by the stripper, which provides for stripping of the pierced material off the punch.

   条料会整个或某个部分卡箍在模具上,通过卸料零件可以将已冲孔的材料从凸模上卸落下来。

5. This punch cuts the blank out of the strip with its outer diameter, forming it afterward with its face area and inner diameter's edge, finally bottoming on a forming support.

   凸模利用外表面的直径部分将落料工作从条料上冲裁下来,然后利用底面面积和内部直径处的刃口边缘对工件进行成型,最终将工作压靠在成型零件的支承面上。

6. A characteristic of compound dies is the inverted position of the blanking die and blanking punch which also functions as the piercing die.

   复合模的典型特征是将落料的凹模和凸模的位置进行倒装,而且落料凸模同时还承担了冲孔凹模的作用。

7. the blanking die is fastened to the upper shoe and the blanking punch having a tapered hole in it and in the lower shoe for slug disposal is mounted on the lower shoe.

   落料凹模安装在上模座,落料凸模安装在下模座,凸模及下模座的内部有一个相互贯通的用于落下废料的锥形阶梯孔。

Lesson 2  Metal Stamping Process and Die Design

## *Check your understanding*

Ⅰ. Give brief answers to the following questions.
1. What is bending?
2. What does a die set consist of?

Ⅱ. Match the items listed in the following two columns.

| | |
|---|---|
| stationary stripper | 下模座 |
| die set | 冲孔加工 |
| piercing | 冲模凹模 |
| side gauges | 模架 |
| lower shoe | 导柱 |
| die buttons | 支承垫板 |
| guide pins | 侧压板 |
| back-up plates | 淬硬工具钢 |
| hardened-tool steel | 固定式卸料板 |

# 金属冲压工艺和模具设计

多种金属冲压技术已经广泛应用在工业和工程中。落料、冲孔、弯曲、拉伸、立体挤压、压花加工、冷挤压是金属冲压技术的一些典型例子。下面讨论其中的一些例子:

弯曲:弯曲工序是一种将坯料弯曲成预定角度的冲压工序。如"L"型托架是将平直的金属板料弯曲呈90°直角。成型工序和弯曲工序的主要区别是弯曲工序形成的是直线状的弯曲(例如盒子的角落),而成型工序可以形成一种曲线状的弯曲。

通常,弯曲需要克服拉应力和压应力。当弯曲完成时,残余应力会使板料回弹至初始位置,因此须将金属板料加大弯曲变形程度以克服残余应力的影响。

落料:落料模在单一工序中通过对工件预期形状的冲裁,产生平板形工件。落料的部分是最终的冲压制件。通常,一套落料模只能冲裁具有外形轮廓而无内部形状要求的工件。

切断:切断模用于切断工件尾部或末端过多的材料,或者是为了后续的工序将毛坯条料切割成预先规定的长度。

拉伸:拉伸工序与成形工序十分相似,不同之处是拉伸工序经历了强烈的塑性变形,工件材料沿着侧面延伸。底部具有细微结构的金属杯是区别拉伸工艺和成型工艺的一个典型实例。金属杯的底部是成型工艺形成的,而四壁是由拉伸工艺形成的。

冲孔:冲孔工序用于形成冲压零件上的孔。

模架是每套冲压模具的基础部分。它包括下模座和上模座,两者加工后呈平行状态,误差要求在几个千分之一英寸之内。有的上模座配有模柄,通过它整个模座和冲床连杆实现夹紧。因为重量偏重,大型冲压模具不会以这种方式安装。它们通过夹具或是螺栓固定在连杆上。但是,有时甚至一些大型模架也可能包含模柄,这种情形用于在冲床上对模具的中心进行定位。

图5.2-1和5.2-2所示的是一套复合模和一套级进模的基本组成部分。两个模座,即上模座和下模座是通过多个导柱实现中心对齐的。导柱在模具工作状态中,为两个模座提供了精确的导向和定位。导柱选用的是经过研磨、渗碳、淬硬后的钢材,它们与下模座呈过盈配合。上模座配置了导套,导柱与导套呈间隙滑动配合。凹模固定板固定在下模座上,它的上面还包括所有的冲模凹模、定位孔和一些弹簧衬垫。凹模固定板也是用淬硬的工具钢加工而成的。

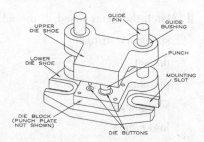

图5.2-1 一套复合模基本组成部分

凹模固定板通常是一块在内部已加工了孔的实心式整体钢或镶拼式组合钢。这些孔必须和模套的外部的形状和直径相匹配:它们的尺寸必须精确,才能保证模套压入后能配合完好。

与凹模固定板相似,凸模固定板以同样的方式安装在上模座上。它也是用一块整体式的或是组合式的淬硬工具钢制成的。在它内部支承了所有的凸模、导向零件、弹簧垫板和模具的其他零件。它们的尺寸和形状应该与相应的配合部分一致,减小压入配合时产生的误差。

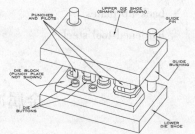

5.2-2 一套级进模的基本组成部分

凸模固定板和凹模固定板通常是用垫板与上下模座分离的。垫板的作用是防止凸模和凹模直接嵌入到硬度较低的模座中。板状金属条料在凹模固定板上表面进料,并在两个导料板之间定位。导料装置有两种类型,即引导条料通过凹模的侧压板和在首次为冲孔或落料工件提供精确定位作用的初始挡料销。条料会整个或某个部分卡箍在模具上,通过卸料零件可以将已冲孔的材料从凸模上卸落下来。卸料板通常是冷轧钢制成的,它上面的孔与凸模是有间隙的。

固定式卸料板安装在凹模固定板的上表面,利用大小相同的螺钉和销与凹模固定板相联接。弹压式卸料板在弹簧力的作用下与固定板存在一定的偏距,它用这种方式与凸模固定板相联接。在倒装冲裁模中,凸模是安装在下模部分而凹模安装在上模部分,所以卸料板也要倒置安装。

图 5.2-3 所示的是一副典型的模具剖视图。图中打料推杆穿过凸模的头部,由弹簧提供给它卸料压力。推杆迫使压力垫块或推件块将力作用在条料上,落料工件在凸模向上运动时被卸下,卸料压力随着凹模的下降而递增。冲孔凹模中也具有类似的一系列杆件,被称为推杆,它们在冲压成型完成后向上运动,使杯状制件脱离凸模的表面。卸料板是固定的,用来阻止剩余的条料在模具打开时,跟随成型/冲裁凸模的运动而作向上运动。凸模利用外表面的直径部分将落料工作从条料上冲裁下来,然后利用底面面积和内部直径处的刃口边缘对工件进行成型,最终将工作压靠在成型零件的支承面上。

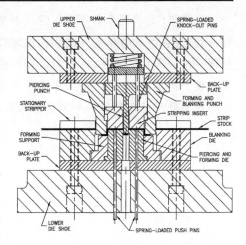

图 5.2-3 一副典型的模具剖视图

冲压模具根据各自的用途,在组成结构上具有很大的差异。一类冲压模具中,进给的金属条料通过模具时,经过多个工步后生产出所要的制件。而另一类模具,通过单一工位的一次冲压就生产出了最终的制件。根据模具的结构和功用,所有冲压模具可分成四种类型:复合模、级进模、钢带模和其他模。

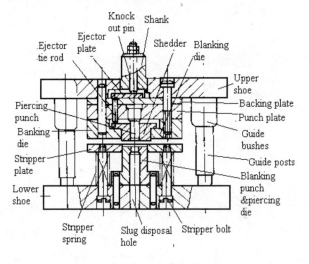

图 5.2-4 复合模

复合模在一个工位下包括了两个或更多的冲压工序。一种复合模可以在一次冲压行程中完成冲孔兼落料的冲裁工序。复合模的典型特征是将落料的凹模和凸模的位置进行倒装,而且落料凸模同时还承担了冲孔凹模的作用。如图 5.2-4 所示,落料凹模安装在上模座,落料凸模安装在下模座,凸模及下模座的内部有一个相互贯通的用于落下废料的锥形阶梯孔。

级进模(见前图 5.2-2)是同一个模架在不同的工位,由许多单个凸模和凹模进行冲裁,组合而成的模具。这些工位定位后,会遵循一定的工序顺序生产,才能制造出所需要的冲压件。通常,凸模和凹模按照并排或水平方向排列。级进模具总体设计比复合模要简单,因为每一道工序并不是完全同时进行。无论进行到哪一步工序,最终的冲压制件只有在最后一个工序结束后才会从条料上分离。

# Lesson 3  Plastics and Injection Molding

*1*  Today the name plastics are identified with the products which are derived from synthetic resins. Plastics are composed of polymer molecules and various additives. As an important class of materials with an extremely wide range of mechanical, physical, and chemical properties, plastics are commonly found in consumer, automotive, electrical and electronic products such as integrated circuits, mechanical equipment, food and beverage containers, packaging, signs, housewares, textiles, safety equipment, toys, appliances, and optical equipment. Plastic materials, commonly called plastics, are well known either as thermosetting or as thermoplastic.

*2*  Plastics are characterized by the following properties: low density, low strength and elastic modulus, low thermal and electrical conductivity, high chemical resistance, and high coefficient of thermal expansion. Also they can be cast, formed, machined, and joined into different shapes and are available in a wide variety of properties, colors, and opacities.

*3*  Many thousands of different plastics ( also called polymers, resins, reinforced plastics, elastomers, etc. ) are processed. Each of the plastics has different melt behavior, product performance and cost. To ensure that the quality of the different plastics meets requirements, tests are conducted on melts as well as molded products. There are basically two types of plastic materials molded: thermoplastics and thermosets. Thermoplastics, which are predominantly used, can go through repeated cycles of heating/melting( usually at least to 260 °C ) and cooling/ solidification. The different thermoplastics have different practical limitations on the number of heating-cooling cycles before appearance and/or properties are affected. As for thermosets, upon the application of initial heat, they also soften and melt like thermoplastics. On the contrary, upon continued application of heating, they undergo a chemical ( cross-linking) change, which hardens them into a permanently hard, insoluble and infusible state. After this they cannot again be softened or melted by further heating.

*4*  There are two main steps in the manufacturing of plastic products. The first is a chemical process to creat the resin. The second is to mix and shape all the material into the finished article or product. Plastic objects are formed by compression, transfer, and injection. Other processes are casting, extrusion and laminating, filament winding, sheet forming, joining, foaming, and machining.

5   Injection molding is a major part of the plastics industry and is the most important process used to manufacture plastic products, consuming approximately 32 wt% of all plastics. It is in second place to extrusion, which consumes approximately 36 wt%. In the United States alone there are about 80,000 IMMs and about 18,000 extruders operating to process all the many different types of plastics. It is ideally suited to manufacture mass produced parts of complex shapes requiring precise dimensions. It is used for numerous products, ranging from boat hulls and lawn chairs, to bottle cups. Car parts, TV and computer housings are injection molded.

6   As shown in Figure 5.3 – 1 the injection molding machine is comprised of:

7   The plasticating and injection unit shown in Fig. 5.3 – 2: The major tasks of the plasticating unit are to melt the polymer, to accumulate the melt in the screw chamber, to inject the melt into the cavity and to maintain the holding pressure during cooling.

8   The clamping unit shown in Fig. 5.3 – 3: It's role is to open and close the mold, and hold the mold tightly to avoid flash during the filling and holding. Clamping can be mechanical and hydraulic.

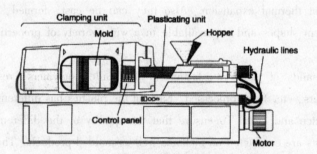

Fig. 5.3 – 1   **An injection molding machine**

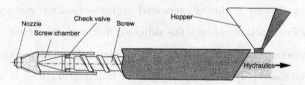

Fig. 5.3 – 2   **Plasticating and injection unit**

# Lesson 3  Plastics and Injection Molding

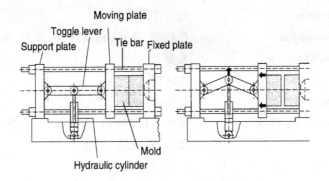

Fig. 5.3 – 3  **Clamping unit**

9  The mold cavity shown in Fig. 5.3 – 4: The mold is the central point in an injection molding machine. Each mold can contain multiple cavities. It distributes polymer melt into and throughout the cavities, shapes the part, cools the melt and ejects the finished product.

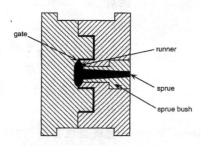

Fig. 5.3 – 4  **Mold cavity**

## ❖ New Words and Phrases

| | |
|---|---|
| synthetic [sin'θetik] a. | 合成的,人造的  n. 人工制品 |
| resin ['rezin] n. | 树脂(松香,树脂状沉淀物,树脂制品) |
| thermosetting [ˌθəːməu'setiŋ] a. | 热硬化性的,热固性的 |
| thermoplastic [ˌθəːmə'plæstik] a. | 热塑性的;塑料热塑的  n. 热塑性塑料 |
| elastic [i'læstik] adj. | 有弹性的 |
| modulus ['mɔdjuləs] n. | 模数 |
| elastic modulus | 弹性模量 |
| thermal ['θəːməl] adj. | 热的,热量的 |
| conductivity [ˌkɔndʌk'tiviti] n. | 传导性,传导率 |
| thermal conductivity | 热传导率 |

| | |
|---|---|
| electrical conductivity | 电传导率 |
| resistance [ri'zistəns] n. | 抵抗力,反抗,反抗行动;阻力,电阻;反对 |
| coefficient [kəui'fiʃnt] n. | 系数 |
| opacity [əu'pæsiti] n. | 不透明 |
| polymer ['pɔlimə] n. | 聚合体 |
| elastomer [i'læstəmə(r)] n. | 弹性体(弹胶物,合成橡胶,高弹体) |
| melt [melt] n. | 熔化,熔化物,熔体 |
| thermoset ['θə:məset] n. | 热固性塑料 |
| cross-linking | 交联 |
| insoluble [in'sɔljubl] a. | 不能溶解的 |
| infusible [in'fju:zəbl] a. | 不能熔化的;难以熔化的;耐热的; |
| compression [kəm'preʃ(ə)n] n. | 压缩 |
| transfer [træns'fə:] v. | 传递 |
| injection [in'dʒekʃən] n. | 注射 |
| extrusion [eks'tru:ʒən] n. | 挤出 |
| laminating ['læmineitiŋ] n. | 层压法(层合法,分成薄层,卷成薄片) |
| filament ['filəmənt] n. | 细丝,细线,单纤维 |
| foaming ['fəumiŋ] n. | 发泡成型,发泡 |
| injection molding | 注射成型 |
| IMMs = injection mold machines | 注射成型设备 |
| extruder [eks'tru:də] n. | 挤出机,挤出设备 |
| plasticating ['plæstikeitiŋ] n. | 塑化 |
| screw [skru:] n. | 螺杆 |
| mold [məuld] n. | 模具 |
| clamp [klæmp] vt. | 夹紧 |
| the clamping unit | 锁模单元 |

### ❖ Notes

1. As an important class of materials with an extremely wide range of mechanical, physical, and chemical properties, plastics are commonly found in consumer, automotive, electrical and electronic products such as integrated circuits, mechanical equipment, food and beverage containers, packaging, signs, housewares, textiles, safety equipment, toys, appliances, and optical equipment.

   作为材料中的一种重要分支,塑料具有非常广泛的机械、物理、化学性能,其产品遍及消费产品、汽车产品、电气和电子产品,如集成电路、机械装备、食品和饮料容器、包装、标记、家用产品、纺织品、安全装置、玩具、器具和光学设备。

**语法补充：with 的用法**

一、本语法在注释一中的应用

　　本注释中介词 with 是具备、拥有的意思，和"an extremely wide range of mechanical, physical, and chemical properties"构成介宾短语，说明了塑料这种重要的材料到底具备什么特性。

二、对本语法的详细阐述

1. 具有；带有

　　如注释一中的"with"即拥有某种特性的意思。

　　e.g. The girl with long hair is my sister.

　　　　那个留长发的姑娘是我妹妹。

2. 和……在一起；由……陪同

　　e.g. You can't see Mary now, as she is with the manager.

　　　　你此刻不到玛丽，因为她正和经理在一起。

3. 表示两件事同时发展。

　　e.g. With time passing by, they have grown up. 随着时间的流逝，他们长大了。

4. 表示伴随某状态一起发生；携带。

　　e.g. The waiter arrived with a cup of coffee. 服务生端着一杯咖啡来了。

5. 引导方式状语

　　e.g. He looked at me with a frown. 他皱着眉看我。

6. 表示原因

　　e.g. The little child trembled with fear. 这个小孩子怕得发抖。

2. Thermoplastics, which are predominantly used, can go through repeated cycles of heating/melting (usually at least to 260 °C) and cooling/ solidification.

　　热塑性塑料是应用最广泛的塑料，它可以反复循环地进行加热/熔融（通常温度高于260°C）和冷却/固化。

3. On the contrary, upon continued application of heating, they undergo a chemical (cross-linking) change, which hardens them into a permanently hard, insoluble and infusible state.

　　然而，再继续进行加热后，热固性塑料将发生化学变化（交联反应）进行固化定型，使其呈现不溶解、不熔融、永久硬化的状态。

4. It is in second place to extrusion, which consumes approximately 36 wt%.

　　除了消耗塑料产量约为36%的挤出成型外，注射成型位居第二位。

5. The major tasks of the plasticating unit are to melt the polymer, to accumulate the melt in the screw chamber, to inject the melt into the cavity and to maintain the holding pressure during cooling.

塑化单元的主要任务是熔融塑料,并将熔体不断地积聚在螺杆头部,然后注射熔体进入模腔,在塑料冷却的过程中保持一定的压力。

## *Check your understanding*

Ⅰ. Give brief answers to the following questions.
1. What are plastics composed of?
2. What are the features of plastics?

Ⅱ. Match the items listed in the following two columns.

| | |
|---|---|
| elastomer | 聚合体 |
| the clamping unit | 注射成型 |
| elastic modulus | 热固性塑料 |
| thermosets | 锁模单元 |
| injection molding | 塑化 |
| polymer | 弹性模量 |
| thermal conductivity | 模具 |
| mold | 热传导率 |
| plasticating | 弹性体 |

## 塑料和注射成型

当今,"塑料"这一名称等同于合成树脂形成的产物。塑料的组成成分是高聚物分子和各种添加剂。作为材料中的一种重要分支,塑料具有非常广泛的机械、物理、化学性能,其产品遍及消费产品、汽车产品,电气和电子产品,如集成电路、机械装备、食品和饮料容器、包装、标牌、家用器皿、纺织品、安全设备、玩具、器具和光学设备。塑料材料,一般称为塑料,大家熟知的名称还有热固性塑料或热塑性塑料。

塑料具有如下特点:密度低、强度和弹性模量小、导电和导热性能低、化学稳定性和热膨胀系数高。而且塑料能进行浇注、成型、机械加工和联接以形成不同的形状,具有范围很广的性能、颜色和透光性。

成千上万种的各类塑料(又称聚合物、树脂、增强塑料、弹性体等)都可以进行加工。每一种塑料都有各自的熔融行为,产品性能和商品价格。为了保证各种塑料的性能符合需求,必须对熔体和成型产品进行检验和测试。可以模塑成型的塑料基本上分成两大类型：热塑性塑料和热固性塑料。热塑性塑料是应用最广泛的塑料，它可以反复循环地进行加热/熔融（通常温度高于260℃）和冷却/固化。每种热塑性塑料的加热-冷却反复循环的次数存在不同的极限，超过极限以后产品的表观质量和性能都会受到影响。对热固性塑料初步加热后，它和热塑性塑料一样会发生软化和熔融。然而，再继续进行加热，热固性塑料将发生化学变化（交联反应），塑料固化定型，呈现不溶解、不熔融、永久硬化的状态。成型后的热固性塑料，若继续加热，它也不能再软化和熔融了。

塑料产品的生产过程分为两个主要阶段。第一个阶段是通过化学过程获得树脂。第二个阶段是将塑料原料通过混合和成型，获得最终的制件或产品。塑料制件成型的方法有压缩成型、传递成型和注射成型，还包括浇注、挤出、层压、抽丝、板材成型、联接、泡沫成型和机械加工。

注射成型是塑料工业的一个主要组成部分，它是生产塑料产品的最重要的加工方法，大约消耗所有塑料产量的32%。除了消耗塑料产量约为36%的挤出成型外，注射成型位居第二位。以美国一个国家为例，大约有80,000台注射机和18,000台挤出机正在运转，加工多种不同类型的塑料。注射成型非常适用于制造大批量的因外形复杂而对尺寸精确有要求的塑料制品。它可以生产各式各样的产品，从船毂、户外椅到水杯。汽车零部件、电视和电脑外壳都是用注射成型制造的。

注射成型设备的构成如图5.3-1所示：

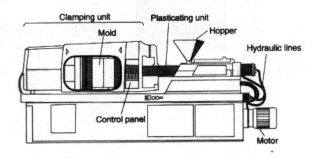

图5.3-1 注射成型设备

塑化和注射单元（图5.3-2）：塑化单元的主要任务是熔融塑料，并将熔体不断地积聚在螺杆头部，然后注射熔体进入模腔，在塑料冷却的过程中保持一定的压力。

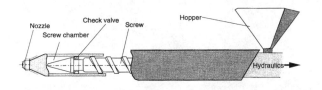

图5.3-2 塑化和注射单元

锁模单元（图5.3-3）：它的任务是开启和闭合模具。它紧紧锁紧闭合后的模具，能够避免在充模和保压过程中出现溢料或飞边。锁模力能够用机械方式和液压方式获得。

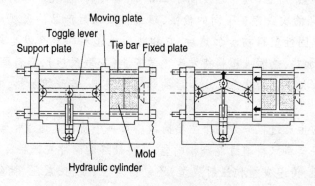

图 5.3-3 锁模单元

模腔单元(图 5.3-4):模具是注射成型设备的主要部分。每套模具上可以拥有多个模腔,塑料熔体进入后充满所有模腔空间,形成各种塑料制件的形状,再对熔体进行冷却,最后推出塑料制品的成品。

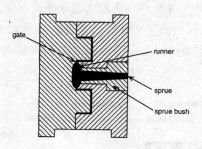

图 5.3-4 模腔单元

# Lesson 4  Design of Injection Mold

*1*  The mold is the most important part of the IMM. It is a controllable, complex, and expensive device. If not properly designed, operated, handled, and maintained, its operation will be costly and inefficient.

*2*  Under pressure, hot melt moves rapidly through the mold. During the injection into the mold, air in the cavity or cavities is released to prevent melt burning and the formation of voids in the product. With PSs, temperature-controlled water circulates in the mold to remove heat; with TSs, electrical heaters are usually used within the mold to provide the additional heat required to solidify the plastic melt in the cavity.

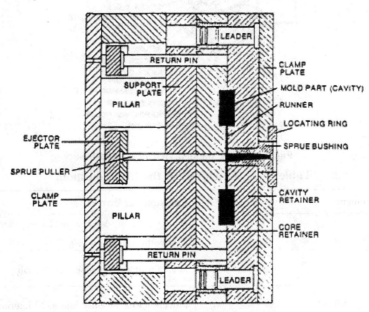

Fig. 5.4 – 1  General configuration of a mold

*3*  The function of a mold is twofold: imparting the desired shape to the plasticized melt and solidifying the injected molded product (cooling for thermoplastics and heating for thermoset plastics). It basically has two sets of components: (1) the cavities and cores and (2) the base in which the cavities and cores are mounted. Figures 5.4 – 1 and 5.4 – 2 and Table 5 – 1 show typical layouts and descriptions of products to be molded that include the cavities and cores. Figure 5.4 – 3 provides an example of the pressure loading of a plastic melt. Melt moves from injection unit, through the mold passageways (sprue, runner, and gate), and into the two cavities.

4  The mold has two basic parts to contain the cavities and cores. They are the stationary mold half on the side where the plastic is injected, and a moving half on the closing or ejector side of the machine. The separation between the two mold halves is called the parting line. In some cases, the cavity is partly in the stationary and partly in the moving section. The term "mold half" does not mean that the two are dimensionally equal in width.

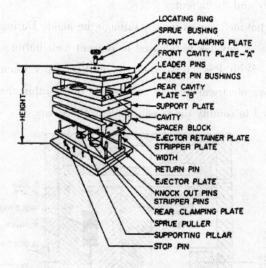

Fig. 5.4 - 2  Exploded view of a mold base

Table 5 - 1  **Functions of the injection mold**

| Mold Component | Function Performed |
| --- | --- |
| Mold base | Hold cavity or cavities in fixed, correct position relative to machine nozzle. |
| Guide pins | Maintain proper alignment of two halves of mold. |
| Sprue bushing (sprue) | Provide means of entry into mold interior. |
| Runners | Convey molten plastic from sprue to cavities. |
| Gates | Control flow into cavities. |
| Cavity (female) and core (male) | Control size, shape, and surface texture of molded article. |
| Water channels | Control temperature of mold surfaces, to chill plastic to rigid state. |

| Side (actuated by cams, gears, or hydraulic cylinders) | Form side holes, slots, undercuts, threaded sections. |
|---|---|
| Vents | Allow escape of trapped air and gas. |
| Ejector mechanism (pins, sleeves, stripper plate) | Eject rigid molded article from cavity or core. |
| Ejector return pins | Return ejector pins to retracted position as mold closes for next cycle. |

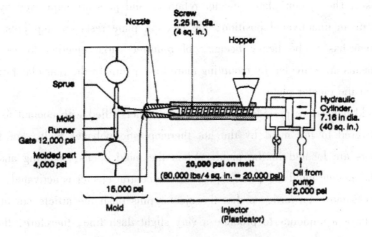

Fig. 5.4-3  **An injector and a mold**

5　The mold determines the size, shape, dimensions, finish, and often the physical properties of the final product. It is filled through a central feed channel, called the sprue. The sprue, which is located in the sprue bushing, is tapered to facilitate mold release. In multicavity molds, it feeds the polymer melt to a runner system, which leads into each mold cavity through a gate.

6　The mold is aligned with the injection cylinder by means of a ring in the stationary mold half. The locating ring surrounds the sprue bushing and is used for locating the mold concentrically with the machine nozzle.

7　The parting line is formed by cavity plates A and B. Cavity plate A retains the cavity inserts and supports the leader pins, which maintain the alignment of cavity halves during operation. These guide pins are preferably mounted in the stationary mold half to ensure that the molded product(s) will fall out of the mold during ejection without being fouled. One of the four leader pins is offset by about 6 in. (0.48 cm) to eliminate the chance of improper assembly of the two halves. On ejector systems, a minimum of four leader pins and bushings are used to prevent cocking of the plate, which reduces wear and prevents seizing.

*8* Mating with plate A is plate B, which holds the opposite half of the cavity or the core and contains the leader-pin bushings for guiding the leader pins. The core establishes the inside configuration of a part. Plate B has its own backup or support plate. The B backup plate is frequently supported by pillars against the U-shaped structure known as the ejector housing. The housing, consisting of the rear clamping plate and spacer blocks, is bolted to the B backup plate, either as separate parts or as a welded unit. This U-shaped structure provides the space for the ejector plate to perform the ejection stroke, also known as the stripper stroke. The ejector plate, ejector retainer, and pins are supported by the return pins. When in an unactivated position, the ejection plate rests on stop pins. When the ejection system has to be heavy because of required large ejection forces, additional supporting means are provided by mounting more leader pins in the rear clamping plate and the bushing in the ejector plate.

*9* Both mold halves are provided with cooling channels filled with coolant to carry away the heat delivered to the mold by the hot thermoplastic polymer melt. For thermosets, electric heaters are located in the mold. When the mold opens, molding and sprue are carried on the moving mold half; subsequently, the central ejector is activated, causing the ejector plates to move forward, so that the ejector pins push the article out of the mold. Ejector pins have a tendency to produce a very slight flash line, therefore, their location and the amount of recess formed by them in the part should be agreed on with the product designer.

### ❖ New Words and Phrases

| | |
|---|---|
| cavity [ˈkæviti] n. | 型腔,模腔,凹模 |
| core [kɔː] n. | 型芯 |
| mold base | 模架,模座 |
| nozzle [ˈnɔzl] n. | 喷嘴 |
| sprue bushing | 主流道衬套 |
| runner [ˈrʌnə(r)] n. | 流道 |
| gate [geit] n. | 浇口 |
| undercut [ˈʌndəkʌt] n. | 侧凹,侧抽芯 |
| sleeve [sliːv] n. | 套筒 |
| stripper plate | 脱料板 |
| return pin | 复位杆 |
| parting line | 分型线 |
| locating ring | 定位圈 |

# Lesson 4  Design of Injection Mold

| | |
|---|---|
| insert [in'sə:t] n. | 镶嵌物,镶件 |
| plate [pleit] n. | 模板 |
| support plate | 支撑板 |
| pillar ['pilə] n. | 柱子 |
| clamping plate | 锁紧模板 |
| ejection stroke | 推出行程 |
| stop pin | 支承挡钉 |

## ❖ Notes

1. The function of a mold is twofold: imparting the desired shape to the plasticized melt and solidifying the injected molded product (cooling for thermoplastics and heating for thermoset plastics).
   注射模具的作用有两个方面:将充分塑化的塑料熔体形成特定的形状,对注射模塑的产品进行固化(热塑性塑料采用冷却方式,热固性塑料采用加热方式)。

2. They are the stationary mold half on the side where the plastic is injected, and a moving half on the closing or ejector side of the machine.
   一半是定模部分,在注射塑料的一侧;另一半是动模部分,在注射设备负责闭合和开启模具的一侧。

3. The housing, consisting of the rear clamping plate and spacer blocks, is bolted to the B backup plate, either as separate parts or as a welded unit.
   由后锁模板和模脚形成的支架结构,既可以是独立零件组合而成,也可以是焊接成整体单元,用螺栓与 B 模板的支撑板联接在一起。

4. When the ejection system has to be heavy because of required large ejection forces, additional supporting means are provided by mounting more leader pins in the rear clamping plate and the bushing in the ejector plate.
   当需要的推出力很大而导致推出机构重量增大时,可提供的额外支撑方法是分别在后锁模板及推板上安装更多的导柱和导套。

**语法补充:because 和 because of 的用法区别**

一、本语法在注释四中的应用

   "because of"后面只能接名词性的成分,如注释四中"required large ejection forces"就是一个名词短语,接在"because of"后面,说明"the ejection system has to be heavy"的原因是什么。

二、对本语法的详细阐述

1) because 是连词,其后接句子;because of 是复合介词,其后接名词、代词、动名词、

what 从句等。

e.g. I didn't buy it because it was too expensive.
我没有买是因为它太贵了。(接句子)

e.g. He is here because of you (that).
他为你(那事)而来这里。(接代词)

e.g. He lost his job because of his age.
由于年龄关系他失去了工作。(接名词)

e.g. We said nothing about it, because of his wife's being there.
因为他妻子在那儿,我们对此只字未提。(接动名词)

e.g. He knew she was crying because of what he had said.
他知道她哭是因为他说的话。(接 what 从句)

2) because 所引导的从句除用作原因状语外,还可用作表语。

e.g. It is because he loves you.
那是因为他爱你。

而只有主语是代词(不是名词)时,才能用 because of 引出的短语作表语。

e.g. It is just because of money.
那只是因为钱的原因。

e.g. That was because of his sickness.
那是因为他生病的原因。

5. Both mold halves are provided with cooling channels filled with coolant to carry away the heat delivered to the mold by the hot thermoplastic polymer melt.
模具的动模和定模两个部分都设置了冷却通道,充满通道的冷却介质将吸收模具从高温的热塑性塑料熔体获得的热量。

## Check your understanding

Ⅰ. Give brief answers to the following questions.
  1. What is the function of a mold?
  2. What two basic parts does the mold have?

Ⅱ. Match the items listed in the following two columns.
  mold base            套筒
  parting line         型芯
  sleeve               主流道衬套
  core                 浇口

| | |
|---|---|
| gate | 型腔,模腔,凹模 |
| cavity | 喷嘴 |
| stripper plate | 模架,模座 |
| sprue bushing | 分型线 |
| nozzle | 脱料板 |

# 注射模具的设计

模具是注射成型设备上最重要的组成部分,它是一套可操纵的、结构复杂且价格昂贵的设备。如果模具设计不合理,或者操作、安装及维护环节不合理,运行模具的成本将很高,而生产效率不高。

在注射压力下,温度很高的塑料熔体迅速地在模具中流动。在熔体注入模具过程中,模具凹模或型腔中的气体要获得释放,才能避免熔体燃烧或在制品中形成气泡。如果是热塑性塑料,温度可调的水在模具内部循环流动移走热量;如果是热固性塑料,通常在模具内部安放电加热器,为固化模腔中的塑料熔体提供所需要的额外热量。

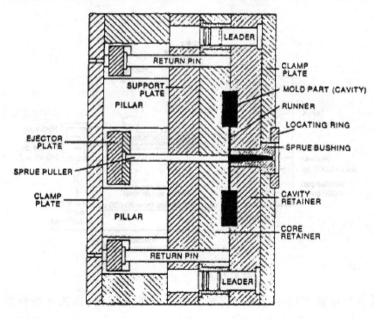

图 5.4-1 注射模具的典型结构

注射模具的作用有两个方面:使充分塑化的塑料熔体形成特定的形状,对注射模塑的产品进行固化(热塑性塑料采用冷却方式,热固性塑料采用加热方式)。它基本上是由两大部件构成:(1)凹模和型芯;(2)安装凹模和型芯的模架。图 5.4-1 和 图

5.4-2所示的是注射模具的典型结构,表格5-1显示的是包括型腔和型芯在内的注射模具的主要组成部分及作用。图5.4-3显示的是塑化后的熔体在压力作用下的一个典型事例。塑料熔体在注射单元作用下流动,通过模具的入口通道(主流道,分流道和浇口)后进入到两个型腔中。

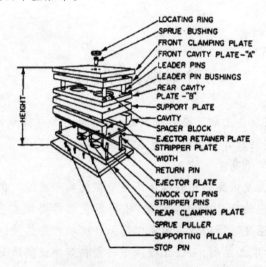

图5.4-2 模架的分解图

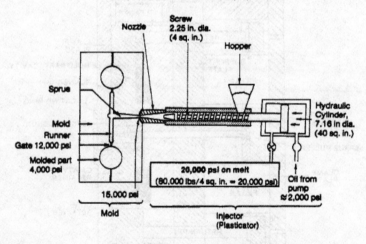

图5.4-3 注射机与模具

注射模具分成两大部分容纳凹模和型芯。一半是定模,在注射塑料的一侧;另一半是动模,在注射设备负责闭合和开启模具的一侧。两半的分离处称为分型线(或分型面)。在某些情况下,模腔会一部分位于定模,另外一部分位于动模。"模具的一半"这个术语并不是指动模和定模宽度完全相等。

注射模具决定了塑料最终产品的规模、形状、尺寸、精度、粗糙度和物理性能。熔体通过中间的一个进料通道,又称主流道,进入模腔。主流道位于浇口套中,形状呈圆

## Lesson 4  Design of Injection Mold

锥体,有助于凝料的脱模。在多腔模中,塑料熔体经过主流道后流入分流道,最终通过浇口进入模具的每一个型腔。

模具与注射机机筒通过定模上的定位圈轴线对齐。定位圈处在浇口套的周围,用于将模具中心与机筒上喷嘴的中心轴线重合。

表5-1  注射模具的组成及作用

| 模具的组成 | 作用 |
| --- | --- |
| 模架 | 安装、固定模具型腔,并准确地与注射机上的喷嘴定位 |
| 导柱 | 使模具定模和动模两个部分完全对中 |
| 主流道衬套 | 熔体流入模具内部的通道 |
| 流道 | 输送熔体从主流道到型腔的通道 |
| 浇口 | 控制熔体进入型腔的流动 |
| 凹模(母模)和型芯(公模) | 成型塑料制品的尺寸、形状和表面质量 |
| 冷却水道 | 控制模具表面温度,将塑料冷却至固体 |
| 侧抽芯 | 形成侧孔、凸台、侧凹和螺纹 |
| 排气槽 | 释放残留在模具型腔中的气体 |
| 推出机构(推杆、推管、推板) | 将固化的成型产品从凹模或型芯上推出 |
| 推出机构复位杆 | 模具闭合时,用来恢复推出机构位置的推杆,为下一个注射成型周期做好准备 |

分型线是构成型腔的 A 模板和 B 模板形成的。A 模板上支承凹模镶件和导柱,它们的作用是保持型腔的两半部分在工作过程中的定位和导向。为了保证成型制品在推出过程中从模具中不受阻碍或污染地脱落下来,比较合理的设计是将导柱安装在定模一侧。四根导柱中的一根其位置必须偏移约 6 英寸(或 0.48 厘米),这样可以消除定模和动模两个部分不正确的装配。在推出机构中,为了阻止推板的倾斜,至少安装四根导柱和导套,从而减小磨损和防止卡住。

与 A 模板匹配的是 B 模板。它的上面固定了与凹模相对的型芯,安装了引导导柱运动的导套。型芯的作用是成型制品的内部结构。B 模板上还设置了一块支撑板。支撑板通常用支撑导柱进行支承,形成众所周知的"U"状支架结构。由后锁模板和模脚形成的支架结构,既可以是独立零件组合而成,也可以是焊接成整体单元,用螺栓与 B 模板的支撑板连接在一起。"U"状结构为推板推出塑件的行程提供了空间。推板、推杆固定板和推杆由复位杆支承在一起。当模具处于非活动状态时,推板停留在支承挡钉上。当需要的推出力很大而导致推出机构重量增大时,可提供的额外支撑方法是分别在后锁模板及推板上安装更多的导柱和导套。

模具的动模和定模两个部分都设置了冷却通道,充满通道的冷却介质将吸收模具从高温的热塑性塑料熔体获得的热量。对于热固性塑料,模具中安装的是电加热装置。模具打开后,成型的制件和流道的凝料被保留在动模的一侧;接着,中间的顶杆开始运动,迫使推板向前运动,从而带动推杆将塑件从模具中推出。推杆的不良后果是形成轻微的飞边,所以,在塑件上设计推杆的位置和推杆形成的凹坑数量应与塑料制品设计人员达成一致意见。

## 构词法之转化法

转化法的构成方式(续)

(四)副词转化为动词

better([副词]更好地)→better([动词]将…变得更好)

e.g. We will try our best to better our living conditions.

我们会尽力改善我们的生活条件。

(五)名词转化为动词

wax([名词]蜡)→ wax([动词]打蜡)

e.g. The surface of this "plaster" may be further refined and may be painted and waxed to resemble a finished bronze casting.

这种"石膏"表面可以进一步完善,着色、打蜡后就近似于青铜铸件了。

(六)形容词转化为动词

average([形容词]平均的)→ average([动词]取平均值)

e.g. Aluminium and magnesium products average about 13.5 kg as a normal limit.

铝、镁产品一般重量限制在平均 13.5 千克。

(七)副词转化为动词

out([副词]出,在外)→ out([动词]暴露)

e.g. Murder will out.

恶事终必将败露。

# Unit 6

## Lesson 1  Food Robotics

*1*  Deploying robotics in many industries is so routine that it is taken as a given. The presence of robotics in other industries, such as in the food market, is relatively low. The potential for robotics in the food and beverage industry is immense, for both "traditional" applications such as picking, packing and palletizing, as well as for cutting-edge applications such as meat cutting and beverage dispensing.

*2*  A lot of food manufacturers are limited because they do not think about robotics. " If food manufacturers use standard robotic systems for the transferring and cutting of their product, they will see more flexibility," says Sylvie Algarra, Product Marketing Manager at Stäubli Robotics, Duncan, South Carolina.

*3*  Robotics in the food and beverage industry are generally divided into three main categories, picking, packing and palletizing. Picking is usually the first of these processes, followed by the packing, then palletizing. Michael Crane, Consumer Goods Segment Manager at QComp Technologies, Inc., Greenville, Wisconsin, defines robotic picking to be "high-speed individual pick and loading trays. Examples are high-speed robots picking chocolates and loading them into a wrapping machine."

*4*  Because picking operations tend to be early in the food manufacturing process, it is important that they be done correctly. Richard Tallian, Consumer Industry Segment Manager, Robotic Products Group at ABB, Inc., speaks of robotic food and beverage picking. "Food robotics have more challenges at the beginning of the production line than at the end, because picking deals with food products that are irregular in shape. When going down the production line, product is more consistent in shape."

Fig. 6.1-1  Meat-cutting application, developed by CRIQ (Industrial Research Center of Quebec)

*5*  Tallian adds that product orientation is something that integrators have to keep in mind with robotic food picking applications. "Generally, product is randomly oriented so some may be touching or overlapping," Tallian says. "Operators have to determine whether they are seeing bad product or overlapping product. Vision has a lot to deal with picking applications." Tallian also says vision systems determine orientation of the product for correct placement.

6    Robotic packing of food is segmented into primary and secondary packaging. Primary packaging is the placing of foods into the first wrapper or layer of packaging. Secondary packaging has the robot inserting the primary wrapped food product into the next layer of packaging, such as a box, case, carton or tray.

7    Laxmi P. Musunur, Packaging Segment Manager at FANUC Robotics America, Inc. speaks of robotic food packaging applications. "Robotics have the most penetration in the packaging side of the food industry. Primary packaging is product that has been wrapped in one form, while secondary packaging has primary packaged foods put into trays or boxes," says Musunur. He continues by saying that a lot of packaging is done by hard automation, but the role of robotics is on the rise. "FANUC has installed fewer robots in the packaging of raw foods, but that is an area where we see significant opportunities."

8    Sylvie Algarra of Stäubli says that robotic food handling often has several applications being undertaken on the same production line, including packaging. "Food is being processed, with other applications, such as slicing, positioning, and dish-making. For example, people are buying outdoor meals like cakes, salads and sandwiches. The food industry has to slice these and package them into individual portions." Algarra asserts that a major difference between primary and secondary packing is the need for speed in the former. "In primary packaging, the production rate is important as it is five times faster than in secondary packaging."

Fig. 6.1 - 2  **Applied Robotics' bag gripper handles bags of food**
(Applied Robotics, Inc.)

9    The variety of food packaging gives robotics an advantage over hard automation. Packaging variety was addressed by Clay Cooper, Engineering Manager at Applied Robotics, Inc., Glenville, New York. "Consumer food packaging has gone through many transformations, from cartons, to stand-up pouches, to foil. Unless a new hard automation system is developed for each new type of packaging, food manufacturers have to resort back to manual systems or use robots," Cooper says. "Robots have the inherent flexibility to handle a variety of packaging." Flexibility is the main reason why robots are going to take over the task of packaging food from hard automation, says Cooper.

*10*   Likewise, Derek Rickard, Distribution Systems Manager at RMT Robotics Ltd., Grimsby, Ontario, Canada, says that the ever-changing varieties of packaging styles and materials poses a challenge for robotic food packing applications. "The packaging itself has a great deal to do with the packing process because package design comes in such a variety. That is great for retailers, but for automation companies, it means dealing with a lot of variation, such as boxes, trays and shrink wrap." RMT's picking applications are suited to deal with such a variety of packaging.

Fig. 6.1 – 3   **A gantry food palletizing work cell**
(**Courtesy RMT Robotics Ltd.**)

*11*   John Holmes, Key Account Manager at PIAB Vacuum Products, Hingham, Massachusetts, is also thinking about the wide variety of food packaging. "Robotics is seeing growth in packaging applications because multiple products are coming down a line, which requires flexibility," Holmes says. "Traditional food packaging lines used to handle one type of packaging, now they have to handle up to five and their production runs are much shorter."

*12*   Palletizing has robots putting the cases or cartons that contain packaged foods onto a shipping pallet. While palletizing is usually relatively straightforward, this application becomes more complex when food manufacturers need to ship mixed load pallets. Joseph Reams, Technical Sales Manager at Schmalz, Inc., Raleigh, North Carolina, explains how mixed product pallets are put together robotically. "Schmalz takes several products and creates mixed SKUs to go from a distribution warehouse to the individual store. We put a layer of one item on a pallet, then another layer on top of that with a different product." Schmalz provides grippers for palletizing.

*13*   Because some foods require low temperatures to prevent spoilage, creating mixed pallets often needs to be done within a cold room. Functioning in low temperatures presents greater demands on equipment, particularly grippers. "When mixing SKUs, grippers need to handle that variety of items. Foam pads can handle a wider range of products," says Joseph Reams.

Fig. 6.1 – 4   **A work cell boxing bread sticks from a moving conveyor**
(**Courtesy Flexicell Inc.**)

"Temperature becomes an issue with foam rubber pad on grippers. When foam pads are brought into a freezing environment, they tend to freeze when exposed to moisture. The foam becomes unresponsive," Reams says. When foam padded grippers are not appropriate, integrators turn to using suction cups.

14   Suction cups are also on the mind of John Holmes of PIAB. Holmes says that vacuum-driven silicone suction cups pose no threat of contaminating food products. "Air-driven vacuum pumps do not require oil and do not generate heat, so are very safe," Holmes says. Silicone suction cups must be food grade, Holmes adds.

15   Palletizing applications in the food and beverage sector make heavy use of vision, particularly when building mixed load pallets. Hans Schouten, Vice President of Marketing and Sales at Flexicell Inc. , Ashland, Virginia, delineates the multiple roles of vision in food palletizing applications. "Vision is used for product recognition, finding orientation, and for inspection. For example, as frozen bread rolls come out of a freezer, some of their dimensions might not meet specifications," Schouten says. "Vision systems help pick up the product, but end-users can also use vision as a secondary quality control inspection."

16   Food and beverage robotics perform tasks other than picking, packing and palletizing. Robotics are increasingly used for applications as diverse as butchering of meat and dispensing beverages. Robotic meat cutting is still a rare application in North America, but is more common in Europe. KUKA Robotics Corp. , Clinton Township, Michigan, is one company that has implemented meat cutting work cells in Europe. William A. Willard, National Accounts Manager for Food and Beverage applications at KUKA comments on the function of robotics in meat processing lines.

"KUKA is working with companies that specialize in food processing in Europe who have applied our robots." Willard explains that after an animal has been processed, robots do the prime cutting. He says that robotic meat cutting is performed in a similar way that a butcher does except that a hog, for instance, is oriented vertically rather than horizontally.

17   Robotic meat cutting, like other food and beverage applications, utilizes vision to accomplish its chore. KUKA's systems employ infrared scanning of pigs and cows prior to robotic butchering. Again, KUKA's William Willard: "KUKA does infrared scanning of

Fig. 6.1 - 5  **A KUKA work cell handling beverage crates**
(Courtesy KUKA Robotics Corp.)

Lesson 1  Food Robotics

the body to make a three-dimensional image of it. Internal algorithms determine where the cut needs to be made," Willard says. "The robot's software is interpolating exactly the carcass's precise position and where the robot's arm has to be to take a certain action."

18  Laxmi Musunur of FANUC also says robotics have a role in cutting meat. "A large piece of meat is on a moving conveyor. A vision system using a laser scanner determines the meat's topography and cuts a perfect slice out of it," Musunur says.

Fig. 6.1-6  **An automated bar system**
(**Courtesy Motoman, Inc.**)

19  "With the availability of integrated vision (robot vision built-in to the robot controller) it's now very easy to incorporate vision into a wide range of applications," added Musunur. FANUC Robotics' has an integrated vision system that can be applied to virtually any picking, packing or palletizing application.

20  On the beverage side of food robots, Motoman Inc., West Carrollton, Ohio, offers several unique RoboBar service robot systems for dispensing beverages. RoboBar is available in high-production, entertainment and a non-alcohol versions each designed to fill a particular market niche. Ron Potter, Motoman's Senior Director of Emerging Markets, describes each of them. "Motoman's RoboBar HP (high production) model is meant for high-volume service bars in places like casinos, cruise ships or airports. It can mix hundreds of recipes of drinks through dispensing guns that pump the liquor and mixes, and can also add ice, if desired." Potter gives an example of a mixed drink order. "If a customer wanted a Long Island iced tea, a server would place the order through a touch screen. RoboBar HP would mix the six ingredients, perfectly portioned, in about twenty seconds."

21  The incentive for service bars and casinos to invest in RoboBar HP stems from the fact it can work around the clock and do the equivalent work of four bartenders. Another version of RoboBar, NA or non-alcohol model, dispenses hot drinks such as coffee, espresso, cappuccino, and lattes. Furthermore, RoboBar NA serves up sodas, fruit juices and other non-alcoholic beverages. Motoman has sold a RoboBar NA to a company in Dubai, where it is being installed in a futuristic office building. Finally, the RoboBar E (entertainment model), pours cocktails directly from liquor bottles and one is in the process of being set up in Harrods Department Store in London, according to Potter. "RoboBar E generally uses one arm to pick up a glass and add ice while the other arm pours or dispenses," says

Potter.

22　A Motoman dual-arm robot is at the heart of all three RoboBar models. The robot is one of Motoman's standard robots that is also used for machine loading applications. Potter says, "Automotive manufacturers are currently using about 1,000 of these robots in Japan and the US." He went on to describe RoboBar's gripper: "The system uses a standard Schunk servo gripper, which is ideal for handling a variety of bottle and glass sizes and shapes." Both robot arms are equipped grippers to manipulate bottles and glasses. Automatic beer bottle decapping is also completed by the robot.

## ❖ New Words and Phrases

| | |
|---|---|
| deploy [di'plɔi] v. | 使用 |
| given ['givn] n. | (推理过程中的)已知事物 |
| potential [pə'tenʃ(ə)l] n. | 潜能，潜力，电压 |
| pick [pik] v. | 分拣 |
| palletize ['pælitaiz] vt. | 码垛堆集，货盘装运 |
| cutting-edge | 前沿 |
| dispense [dis'pens] v. | 分发，分配 |
| wrap [ræp] vt. | 包，裹 |
| wrapping machine | 包装机 |
| keep in mind | 牢记 |
| touch [tʌtʃ] v. | 粘连 |
| overlap ['əuvə'læp] v. | 重叠 |
| bad product | 劣质产品 |
| primary packaging | 首次包装 |
| secondary packaging | 二次包装 |
| automation [,ɔːtə'meiʃən] n. | 自动化 |
| hard automation | 刚性自动化装置 |
| raw [rɔː] adj. | 生的 |
| raw food | 生食食物 |
| slice [slais] n. | 切片 |
| dish-making | 装盘 |
| outdoor meals | 室外食品 |
| pouch [pautʃ] n. | 小袋 |
| stand-up pouch | 立袋 |
| foil [fɔil] n. | 铝箔 |

# Lesson 1  Food Robotics

| | |
|---|---|
| resort to | 诉诸,采取 |
| retailer [riːˈteilə] n. | 零售商 |
| shrink wrap | 热缩塑料包 |
| Key Account Manager | 大客户经理 |
| SKU (Stock Keeping Unit) | 包装件 |
| suction [ˈsʌkʃən] n. | 吸,吸入 |
| suction cup | 吸盘 |
| vacuum [ˈvækjuəm] n. | 真空 |
| pump [pʌmp] n. | 抽水机 |
| vacuum pump | 真空泵 |
| food grade | 食品等级 |
| make use of | 使用 |
| work cell | 单元式生产 |
| hog [hɔg; (US) hɔːg] n. | 肥猪 |
| chore [tʃɔː] n. | 家务杂事 |
| infrared [ˈinfrəˈred] n. | 红外线 |
| carcass [ˈkɑːkəs] n. | (屠宰后)畜体 |
| topography [təˈpɔgrəfi] n. | 地形,地势,地貌 |
| roboBar | 机器人酒保 |
| niche [nitʃ] n. | 适当的位置 |
| casino [kəˈsiːnəu] n. | 赌博娱乐场 |
| version [ˈvəːʃən] n. | 形式,种类 |
| around the clock | 昼夜不停,连续一整天 |
| bartender [ˈbɑːˌtendə] n. | 酒吧间男招待 |
| espresso [ˈespresəu] n. | (蒸汽加压煮出的)浓咖啡 |
| cappuccino [kæpʊˈtʃiːnəu] n. | 热牛奶咖啡 |
| latte [lʌti] n. | 拿铁咖啡 |
| dual [ˈdjuː(ː)əl] adj. | 双重的 |
| dual-arm | 双臂 |
| servo [ˈsəːvəu] n. | 伺服 |
| gripper [ˈgripə] n. | 夹子(抓爪器) |
| servo gripper | 伺服手爪 |

## ❖ Notes

1. The potential for robotics in the food and beverage industry is immense, for both

"traditional" applications such as picking, packing and palletizing, as well as for cutting-edge applications such as meat cutting and beverage dispensing.

机器人在食品和饮料行业的应用潜力是巨大的,如分拣,包装和码垛这些传统的应用,以及类切割和饮料配送等较前沿的应用。

2. "Robotics have the most penetration in the packaging side of the food industry. Primary packaging is product that has been wrapped in one form, while secondary packaging has primary packaged foods put into trays or boxes," says Musunur.

Musunur 说:"机器人技术主要渗透到食品工业的包装领域,首次包装是把食品包装成一种形状,而二次包装是把首次包装好的食品放入托盘或箱子。"

3. KUKA's systems employs infrared scanning of pigs and cows prior to robotic butchering.

酷卡的机器人系统在对猪和牛进行宰割之前要使用红外扫描。

4. "With the availability of integrated vision (robot vision built-in to the robot controller) it's now very easy to incorporate vision into a wide range of applications," added Musunur. ?

拉克西米补充说,"随着综合视觉技术的实用(把机器人视觉嵌入到机器人控制器中),很容易就可以把视觉系统应用到许多领域。"

5. The incentive for service bars and casinos to invest in RoboBar HP stems from the fact it can work around the clock and do the equivalent work of four bartenders.

酒吧和赌场投资 RoboBar HP 的动机源于这样一个事实,它可夜以继日的运作并能完成四位酒吧招待的工作。

## 语法补充:同位语从句的用法

一、本语法在注释五中的应用

　　注释五中"that it can work around the clock and do the equivalent work of four bartenders"是一个由关系代词 that 引导的同位语从句,对先行词 fact 进行解释说明,告诉读者这个 fact 的具体内容是什么。

二、对本语法的详细阐述

(一) 定义:同位语从句的作用是对先行词进行解释说明。

(二) that 引导的同位语从句与定语从句的区别。

1. 在同位语从句中,that 只起连接作用,去掉 that,同位语从句的结构和意义是完整的;

2. 在定语从句中,that 充当主语、宾语或表语,若去掉 that,余下的定语从句部分的结构和意义都不完整。

e.g. The news that our football team won the match was exciting. (同位语从句,that 只起连接作用)

我们足球队赢得比赛的消息很令人兴奋。

e.g. The news (that) we heard on the radio was not true. (定语从句,that 作 hear 的宾语)

我们从收音机上听到的消息不是真的。

## Check your understanding

Ⅰ. Give brief answers to the following questions.
1. What are the applications of robotics in the food and beverage industry?
2. What is palletizing?

Ⅱ. Match the items listed in the following two columns.

dispense          包,裹
retailer          分发,分配
automation        夹子
wrap              机器人酒保
raw food          零售商
gripper           装盘
dish-making       自动化
roboBar           生食食物

# 食品机器人技术

在许多行业使用机器人是一件司空见惯、众所周知的事。但是在其他行业如食品市场使用机器人还是相对较少的。机器人在食品和饮料行业的应用潜力是巨大的,如分拣,包装和码垛这些"传统"的应用,以及肉类切割和饮料配送等较前沿的应用。

很多食品制造商的发展受到限制,因为没考虑使用机器人技术。"如果食品制造商使用一流的机器人系统传送和切割他们的产品,他们将看到更多的灵活性。"南卡罗琳娜州邓肯市史陶比机器人公司产品销售经理西尔维娅·阿尔加拉说。

机器人技术在食品和饮料行业大致分为

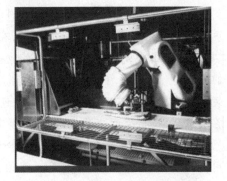

图 6.1-1  由 **CRIQ** 研制的切肉机器人
(魁北克的工业研究中心)

三大类,分拣、包装和码垛。分拣通常是第一个步骤,其次是包装,然后码垛。Qcomp技术公司消费品部门经理迈克尔·克瑞恩认为机器人分拣就是高速的独立分拣和装盘,实例就是高速机器人分拣巧克力并把他们装进包装机。

由于分拣操作往往出现在食品生产的早期,正确的分拣就显得十分重要。ABB公司机器人产品部消费者产业部经理理查德·塔利安谈到机器人进行食品和饮料包装时说:"食品机器人技术在生产线的开始阶段比结束阶段更具有挑战性。因为分拣阶段食品的形状不规则,而顺着生产线向下走,产品在形状上越来越规则。"

塔利安补充说,产品取向是综合者使用机器人分拣食品时必须牢记的事情。"通常产品是都是随便放置,因此有些产品可能粘在一起或者重叠起来。"塔利安说:"操作者必须确定他们看到的是坏的产品还是重叠起来的产品。视觉技术应用于食品分拣时有很多事情要处理。"塔尔利安也表示视觉系统解决了产品的正确安置问题。

机器人进行食品包装分为首次包装和二次包装。首次包装是对食品进行第一次或第一层包装。二次包装就是让机器人把首次包装好的食品放进下一层包装,如箱子、盒子、纸盒或托盘等。

谈到机器人在食品包装中的应用,法鲁克机器人美洲公司包装工程部经理 L. P. Musunur 说:"机器人技术主要渗透到食品工业的包装领域,首次包装是把食品包装成一种形状,而二次包装是把首次包装好的食品放入托盘或箱子。"他继续说,很多包装还都是由刚性的自动化装置完成的,但是机器人的作用正在上升。"法鲁克机器人技术在生食食品包装领域的应用还很少,但是我们却在这个领域看到了巨大的商机。"

图6.1-2 可以处理多种食品袋子的
应用机器人
(应用机器人公司提供)

史陶比机器人公司西尔维娅·阿尔加拉说机器人技术在应用到食品加工时,在同一条生产线上可承担几种应用,包括包装。"在食品加工中,机器人技术还有其他一些应用,例如切片,放置和装盘。例如人们买的室外饭食,象蛋糕、沙拉和三明治。食品工业必须把它们切片,并且进行独立包装。"阿尔加拉断言,首次包装和二次包装之间的主要区别是前者需要速度快些。"在首次包装期间,生产速度是重要的,事实上它比二次包装快五倍。"

食品包装的多样化使机器人操作优于刚性自动化装置。位于纽约格兰威尔的应用机器人公司工程部经理克莱·库伯论述了食品包装的多样性。"消费食品的包装已经经过了多种变革,从纸箱,到立袋,再到铝箔。除非为每一种新包装都开发一个新的刚性自动化装置,否则食品制造商必须采用手工操作或机器人,"库伯说。"机器人在处理各种包装方面具有与生俱来的灵活性。"库伯指出,灵活性是机器人从刚性自

# Lesson 1　Food Robotics

动化装置接管包装食品任务的主要理由。

同样,加拿大渥太华格林姆斯比市 RMT 机器人技术有限公司配送系统经理德里克·里卡德说,品种不断翻新的包装样式和材料给机器人应用于食品包装提出了挑战。"在包装过程中,包装本身有大量工作要做,因为包装设计品种如此之多。这种情况对零售商来说是很好的,但对自动化公司来说,意味着要应付更多的包装形式,如箱子,托盘和热缩塑料包。"RMT 公司的分拣应用技术可以应付这种多样的包装。

图 6.1-3　食品码垛的工作单元
(**RMT** 机器人有限公司提供)

麻省海厄姆市 PIAB 真空产品大客户经理约翰·霍姆斯也在思考食品包装的多样性。"机器人技术在包装应用中的增长,是因为生产线上出现了多种食品,这需要灵活性,"霍姆斯说。"传统的食品包装生产线只能处理一种类型的包装,现在他们必须处理五种并且生产周期要短得多。"

码垛就是让机器人把装有包装好的食品箱子或纸盒子放到货盘上。虽然码垛通常是相对简单的,但当食品制造商需要运送混杂在一起的食品托盘时,此过程就会变得更加复杂。施马尔茨公司技术销售经理约瑟夫·瑞姆解释了如何用机器人把多种产品托盘放在一起。"施马尔茨把几种食品拿来,制作成混合包装件,放到配送货栈,我们把摆放一种产品的托盘置于下部,上面再摆放另

图 6.1-4　面包条装盒的
工作单元
(**Fiexicell** 有限公司提供)

一种产品,再送到一个一个的商店去。"施马尔茨公司提供码垛用的抓爪。

由于一些食物要求低温防止损坏,因此制作混合货盘时常需要在冷室之内进行。在低温下操作给设备提出较高的要求,特别是抓爪。

"在制作混合包装件时,抓爪需要接触多种食品。泡沫垫可以处理更多种类的食品。"约瑟夫·瑞姆说。"温度就成为泡沫垫抓爪的一个问题。泡沫垫在结冰的环境里,遇潮就会结冰,泡沫就不起作用了。"瑞姆说。当泡沫垫做的抓爪不适合使用时,包装商可使用吸盘。

吸盘也是 PIAB 公司约翰·霍姆斯考虑的问题。霍姆斯说,由真空驱动的硅酮吸盘不会造成食品污染。他说:"空气驱动的真空泵不需要油,不产生热,所以是非常安全的。"霍姆斯补充说,硅树脂吸盘必须是食品级的。

在食品和饮料部门应用机器人码垛需要大量使用视觉技术,特别是当制作混装托

盘时。Flexicell 公司行销和销售副总裁汉斯·舒霍顿这样描述视觉技术在食物货盘应用中的多种作用。"视觉技术可用于产品的识别、定位和检测。例如,冻结的小圆面包从冷冻机里滚出来,一些产品尺寸可能不符合要求。"舒霍顿说,"视觉系统有助于把掉了的食品捡起来,最终用户还可以使用视觉技术进行二次质量检查。"

食品和饮料机器人除了执行分拣、包装和码垛的任务之外还可以执行别的任务。机器人技术的应用领域越来越多,如割肉和配送饮料。机器人切肉在北美洲仍然是一种罕见的应用,但是在欧洲却比较常见。密执安克林顿小镇的酷卡机器人公司已经在欧洲实现了机器人进行肉切割的单元式生产。酷卡公司食品和饮料技术应用部全国客户经理威廉·维拉德对机器人在肉加工生产线中的应用发表了评论。"与酷卡共事的公司是那些在欧洲专门进行食品加工且使用了酷卡机器人的公司。"维拉德解释说,在动物被初步加

图 6.1-5　处理饮料箱的工作单元
(KUKA 机器人公司提供)

工之后,机器人做了大部分切割工作。机器人切割肉的方式和屠夫是一样的,只是肉猪等被垂直安置而不是水平放置。

机器人切肉,就像在其他食物和饮料中的应用一样,都运用视觉技术来完成它的差事。酷卡的机器人系统在对猪和牛进行宰割之前要使用红外扫描。威廉·维尔德说:"酷卡对工件进行红外扫描后形成一个三维图象。内部算法确定需要在哪里进行切割,软件确定牲畜的准确位置,并确定机器人的手臂应该在什么位置进行切割。"

法鲁克公司的拉克西米也讲了机器人在肉切割中的作用。"一大片肉放在移动的传送带上,视觉系统用激光扫描器确定它的位置后,就可切出一块完美的肉片。"

拉克西米补充说,"随着综合视觉技术的实用(把机器人视觉嵌入到机器人控制器中),很容易就可以把视觉系统应用到许多领域。"法鲁克机器人公司就有这样一个机器人综合视觉系统可应用于任意分拣、包装和码垛。

针对食品机器人应用于饮料,美国 Motoman 公司提供了几个独特的名为 RoboBar 的服务系统机器人,用于配送饮料。RoboBar 有批量的、娱乐的和无酒精几款型号,每个型号都针对独特的市场空间。Motoman 公司新兴市场部资深主管波特就以上产品一一进行说明。"Motoman 公司的 RoboBar HP(批量生产款)可以为销售大量饮料的酒吧服务,如赌场、游轮或者机场。它用配送枪把抽出来的液体混合,可调制成数百种饮料,如果需要的话还可添加冰块。"波特举了一个例子,来说明它的配制顺序。"如果顾客想要一份长岛冰茶,服务器将通过触摸屏发出订单,RoboBar 就会在约二十秒的时间内按照正确比例把六种配料混合好。"

酒吧和赌场投资 RoboBar HP 的动机源于这样一个事实，它可夜以继日的运作并能完成四位酒吧招待的工作。另外一款 RoboBar NA（无酒精款），可配送热饮，例如咖啡、浓咖啡、热奶咖啡和拿铁咖啡。此外还可提供汽水、果汁和其他无酒精饮料。Motoman 已经卖了一款 RoboBar NA 给迪拜的一家公司，安装在一座未来派的办公大楼里。最后是 RoboBar E（娱乐款），直接地从酒瓶里倒鸡尾酒，伦敦 Harrods 百货商店正在安装一个。罗伯特说，"RoboBar E 通常用一手端酒杯并加冰，另一只手倒酒或送酒。"

图 6.1-6　自动服务的酒吧系统
（**MOTO** 公司提供）

Motoman 双臂机器人位于三款 RoboBar 机器人酒吧服务系统的中心。该机器人是一个 Motoman 标准机器人，也可用于装货。波特说："目前在美国和日本，汽车制造商大约使用着 1,000 个这样的机器人。"他这样描述 RoboBar 的抓爪"系统使用了一个标准的熊克伺服抓爪，用来处理各种大小和形状的酒瓶和杯子是非常理想的。"两条机器人手臂上都装了这种爪子，可操作瓶子和杯子。该机器人还可自动去掉啤酒瓶的瓶盖。

## Lesson 2  Robotic Inspection

*1*  Robot-based inspections systems are an application whose time has come. As vision systems become increasingly powerful and flexible, more end-users will consider inspection tasks being integrated into robotic work cells. Robot makers and integrators can offer end-users some valuable advice on having vision systems do more than just guide the robot.

*2*  "Robotic inspection systems are performing flaw detection on parts, ensuring complete part assembly, and measuring parts," asserts, Eichler, Director of Marketing for Vision Systems at Cognex Corp. "The vision system must be able to both find and inspect the part accurately. Most importantly, integrators have to make sure of getting very good positional accuracy and communicating that back to the robot quickly."

**Parts or No Parts?**

*3*  Inspection systems are called upon to determine part presence. "Integrators start by looking to see if certain things are present or not present on an assembly," says Steve, Business Development Manager for Robotics Vision Technologies "Inspection systems could be looking at an engine to confirm that it has been completely assembled. For example, at the end of the production line, car makers want to confirm that an oil filter has been put on the engine or determine if a certain bolt has been tightened down completely."

Fig. 6.2 – 1  Vision-equipped robot locating and picking randomly-oriented parts
( Courtesy FANUC Robotics American Inc. )

Fig. 6.2 – 2  Inspection tool mounted on robot arm performing inspection of a crankshaft
( Courtesy ABB Inc. )

# Lesson 2　Robotic Inspection

4　The robotic form of "go/no go" inspection utilizes a camera mounted on the robot's arm, which is moved around to check the presence of different features on a part.

**Measure Up**

5　Robots are also used to measure items. "Inspection systems are measuring components but as tolerances of the measurements get tighter and tighter, these tolerances become harder to satisfy," points out Roney, Development Manager of Intelligent Robotics with FANUC Robotics America Inc., "Lighting and part presentation to the robot becomes more critical. When moving from verifying a part's presence to actually measuring it, integrators are adding complexity to the inspection system."

6　As an example of robotic inspection, Roney says the vision system is ascertaining if a nut or a bolt is where it should be or that a hole is tapped properly. "Those features are typically inspected with robotics," says Roney.

7　"SHAFI does three-dimensional robotic metrology that requires dimensional tolerance checking," says Shafi, President of SHAFI Inc. "The robot's inspection system acts as a coordinate measuring machine so end-users do not need to send a sample of parts to a laboratory to check their measurements." Shafi adds that robotic inspection ensures that end-users are given correct answers when collecting data on their parts.

**Error-Proofing**

8　"Robotic flaw detection is looking at surface finishes or finding precise dimensions," describes Eichler of Cognex. "End-users must define what is a good part and what is a bad part. If you ask several line operators to define what is a good part and what is a bad part and they do not all give the same answer, the inspection system will also struggle with finding a good answer."

9　Knowing details of an inspection application is crucial to achieving success. "Issues relating to the detail of the part and the inspection requirements are important," stresses Garmann, Software and Controls Technology Leader at Motoman Inc. "Integrators must know the specifics of the inspection requirements, such as the need to identify an extremely fine defect or detail, before selecting the inspection equipment to fit the application."

10　Likewise, notes Steve Forrest, Applications Manager, at Autotool Inc., Plain City, Ohio, "End-users have to define what the tolerance on the part is and define the appropriate way to do the inspection." To meet these challenges, integrators must use the right inspection method, professes Forrest.

11　Ed Roney of FANUC also emphasizes the importance of distinguishing good parts from flawed parts. "End-users need to understand exactly what needs to be verified or measured. Those doing robotic inspection not only need to know what the good state of an item is but also the bad state. When a bad part shows up, the system has to be configured to properly deal with it."

Fig. 6.2 – 3  **Arobot inspecting automotive parts**
( **Courtesy Autotool Inc.** )

12　Roney adds that when setting up an inspection system, integrators need all the information from the end-user of what constitutes both good and bad parts. "Integrators often get a sample set of parts that might not be a true representation of what is going through long term," Roney cautions.

13　Adil Shafi contends that robotic inspection systems offer cost savings over traditional inspection solutions. "Robotic inspection improves quality in manufacturing because robots can do inspection on every part rather than just on samples." Traditional quality inspection has only one or two percent of parts sent to a laboratory to be checked out. "Quality and cost pressures drive robotic inspection," maintains Shafi.

14　Inspecting all parts rather than just a small sample is important to ABB's end-users, observes Steve. "Traditional inspection has some parts shipped to the quality department, where these parts are put into a coordinate measuring machine for a quality check. With robotic inspection, manufacturers can perform in-line quality checks. The advantage is that manufacturers can check every part rather than just one out of 100."

15　Being able to perform in-line inspection saves on scrap, time and money, says Lisa Eichler. "Robotic inspection adds quality checks into the manufacturing process earlier, so end-users can stop adding value to a bad part and gets it off the assembly line sooner and more reliably." Eichler goes on to say while inspection adds complexity to vision applications in addition to just locating a part, the benefit stems from cost-savings through increased quality.

## Inspecting Experience Counts

16　While inspection systems are getting more powerful and user-friendly, learning from those who are experienced can save time and money. Getting an experienced integrator tops the to-do list for end-users who need to get their vision systems up to speed quickly. Also, end-users can save time and money by doing as much research on the capabilities of inspection systems as is practical.

17　"Having an experienced integrator and getting good technical support are important.

## Lesson 2  Robotic Inspection

In challenging applications such as random bin picking, end-users need to have an integrator who has thorough knowledge in image processing, programming, advanced math, as well as robotics," advises KJ Fredriksson, Integration Manager, with SICK Inc. "North America has about 1,000 vision integrators, and only approximately 30 are certified or authorized vision integrators for SICK."

Fig. 6.2-4  **Bin picking operation done with a robot-mounted 3-dimensional camera** (**Courtesy SICK Inc.**)

18    Andrew Valentine, President and Chief Executive Officer of Valentine Robotics Inc. recommends, "The absolute first step is that end-users have to get good training and knowledge on the robot they are working with and on the inspection technology itself. If problems arise, end-users need to be able to differentiate if the issue is the robot or the inspection system."

19    ABB's Steve West suggests to end-users of robotic inspection systems, "Take time to define the requirements of the application up front. Also, avoid getting too complex with the inspection solution and make sure that end-users have a strategy to anticipate, respond to, and manage false rejects."

20    Adil Shafi also stresses the need to find an experienced vision integrator. "End-users should do their homework. That way, users of inspection systems will be aware of how the technology works and what to expect of the solution and potential results. This ensures the inspection system fits their needs." Shafi emphasizes that those purchasing an inspection system for the first time should know what the dimensional tolerances of their applications are.

21    "Sometimes it takes a demonstration or a test to alleviate end-users' fears and to ensure the inspection system will work. Also, first-time users should request a tour of a plant where the vendor's system is already working," proposes Adil Shafi. "That ensures that the inspection vendor knows what they are doing, and can show the end-user what they are going to get." Shafi also mentions that the timing of inspection images is important and advises that inspections be done when sparks are not being generated during welding inspection applications. Otherwise, the flash of light will cause problems in the vision system's ability to capture a clear image of the part being welded.

22    FANUC Robotics' Ed Roney also believes that an integrator's experience is paramount. "Any vendor can sell an inspection system but only experienced vendors can apply it correctly and understand the issues involved," Roney says. "If an end-user has a

system integrator implement the inspection, the end-user should look at the experience of that integrator. If the end-user is going to integrate the inspection system themselves, they need to do more homework and investigate what is required for their application."

23    Motoman's Greg Garmann urges end-users to, "Clearly identify the scope of the inspection and select inspection equipment to have higher specifications than required. This will provide some flexibility to enhance the inspection if needed or to overcome unforeseen obstacles."

**Inspections to Come**

24    As vision systems become more flexible and powerful, inspection systems will be able to better deal with environmental factors such as coping with the robot's changing dimensions as it warms during operation. Adil Shafi explains, saying, "In the next few years, inspection systems will be more accurate and will have better thermal compensation techniques. When the robot heats up, a millimeter expansion creates a significant change in inspection reporting results." Shafi also sees more three-dimensional vision packages offered in the market.

25    Greg Garmann anticipates that advancements in vision and inspection equipment will continue to simplify the inspection tools and process. "This trend will continue as industry continues to require more inspection data while consumers demand for quality manufactured products increases."

26    Ed Roney foresees greater acceptance of vision-based inspection systems. "I see a trend of more people becoming comfortable with the use of inspection systems. More companies are using vision for robot guidance as well as for error-proofing. They are finding value in vision systems more so than ever before."

27    If a vision system is already deployed for robot guidance, more end-users are thinking that the system can inspect the work the robot is doing, says Steve West. "More people are looking at how to use a robot for inspection. If the robot is already carrying out material handling tasks, not a lot of additional cost is necessary to use the same robot to do inspection by simply adding an additional sensor."

28    KJ Fredriksson of SICK also sees a transition from two-dimensional to three-dimensional inspection. "Today, SICK does 3D Vision primarily by triangulation." "In the next five years, faster computers will lead to more stereo vision inspection." Computers are not fast enough yet and stereo capable vision requires faster processors to do all the math in cycle times required for manufacturing. Increased computer processing power will open doors to stereo vision inspection. Fredriksson adds, "We know how to do stereo-based

# Lesson 2  Robotic Inspection

acquisition today but the bottleneck is computer processing power."

29    In the coming years, robotic inspection will enter the mainstream so most manufacturers will be able to routinely implement it. Also, look for vision systems with the ability to inspect multiple items simultaneously.

## ❖ New Words and Phrases

| | |
|---|---|
| work cell | 单元式生产 |
| flaw [flɔː] n. | 裂纹,瑕疵 |
| detection [diˈtekʃən] n. | 探测 |
| flaw detection | 探伤检验,裂缝检查 |
| complete [kəmˈpliːt] a. | 全部的 |
| make sure of | 确保 |
| call upon | 要求 |
| determine [diˈtəːmin] v. | 测定 |
| filter [ˈfiltə] n. | 过滤器 |
| oil filter | 滤油器 |
| presentation [ˌprezenˈteiʃən] n. | 描述,表示 |
| nut [nʌt] n. | 螺母 |
| bolt [bəult] n. | 螺栓 |
| tap [tæp] n. | 攻螺纹 |
| metrology [miˈtrɔlədʒi] n. | 测量 |
| fine [fain] a. | 细微的 |
| cost savings | 节省成本 |
| check out | 检查 |
| top [tɔp] v. | 达到顶端 |
| to-do list | 待办事项 |
| challenging [ˈtʃælindʒiŋ] a. | 复杂的 |
| up front | 预先 |
| error-proofing | 纠错 |
| material handling | 物料输送 |
| bottleneck [ˈbɔtlˌnek] n. | 瓶颈 |
| mainstream [ˈmeinstriːm] n. | 主流 |

## ❖ Notes

1. Robot makers and integrators can offer end-users some valuable advice on having vision

systems do more than just guide the robot.

机器人制造商和综合者可以为终端用户提供一些宝贵的建议,使视觉系统的应用不只局限于机器人的引导工作。

## 语法补充:使役动词 have 的用法

一、本语法在注释一中的应用

注释一中"having vision systems do more than just guide the robot"用了 have sb/sth do sth 这种搭配,意思是"让某人或某物做某事"。

二、对本语法的详细阐述

1. have + 宾语 + 不定式(不带 to)

这种结构一般表示"让某人做某事"。

e.g. I should like to have you meet Mr. Davis.

我想让你和戴维斯先生认识一下。

2. have + 宾语 + 过去分词

这种结构一般有两种意义,一为"致使",二为被动。前者表示主语的意志致使某事发生或被做到,后者则与主观意志完全无关。

1)表示主观的意志

e.g. I must have this table photocopied.

我必须(找人)把这个表格复印出来。

2)与主观意志完全无关

e.g. I had my pocket picked on the subway.

我在地铁里,口袋被人掏了。

3. have + 宾语 + 现在分词

这种结构有两种意思,一是使某人正在做某事,二是与否定词连用,表示不许某人做某事。

1) have + 宾语 + 现在分词(叫某人做某事)

e.g. We finally managed to have her talking about herself.

我们终于使她开始谈自己的事。

2) not have + 宾语 + 现在分词(不许某人做某事)

e.g. I can't have you smoking in the sitting room.

我不能让你在起居室里抽烟。

2. For example, at the end of the production line, car makers want to confirm that an oil filter has been put on the engine or determine if a certain bolt has been tightened down completely.

# Lesson 2   Robotic Inspection

例如,在生产线的末端,汽车制造商需要确认滤油器是否安装在引擎上或螺栓是否完全拧紧。

3. As an example of robotic inspection, Roney says the vision system is ascertaining if a nut or a bolt is where it should be or that a hole is tapped properly.

作为机器人检测的一个例子,罗尼说,视觉系统可以确定一个螺母或螺栓是否安装到位[3],孔是否攻了螺纹。

4. Those doing robotic inspection not only need to know what the good state of an item is but also the bad state.

做机器人检测的工程人员不仅需要了解项目处于良好状态时是什么样的,还需要了解项目处于不正常状态时又是什么样子的。

5. Robotic inspection adds quality checks into the manufacturing process earlier, so end-users can stop adding value to a bad part and gets it off the assembly line sooner and more reliably.

机器人检测技术使得质量检查可以较早地介入到生产过程中,这样最终用户就可以停止在不合格零件上继续增加价值,并且较早地、更为可靠地把它从装配线上撤下来。

6. Inspection systems will be able to better deal with environmental factors such as coping with the robot's changing dimensions as it warms during operation.

因此检测系统将能够更好地处理一些由于环境因素造成的问题,比如在机器人操作过程中,由于温度升高会导致其结构尺寸有所变化。

7. If the robot is already carrying out material handling tasks, not a lot of additional cost is necessary to use the same robot to do inspection by simply adding an additional sensor.

如果已经使用了机器人进行物料输送任务,只需另外增加一个传感器就可以使机器人进行检测工作,而不必增加额外的费用。

## Check your understanding

Ⅰ. Give brief answers to the following questions.

1. Compared with traditional inspection solutions, what advantages do robotic inspection systems have?

Ⅱ. Match the items listed in the following two columns.

| work cell | 螺栓 |
| nut | 裂缝检查 |
| flaw detection | 视觉系统 |
| material handling | 单元式生产 |

oil filter　　　　　　　　　　　螺母
error-proofing　　　　　　　　滤油器
bolt　　　　　　　　　　　　　纠错
vision system　　　　　　　　　物料输送

# 机器人检测

机器人检测系统的应用时代已经到来。因为视觉系统变得越来越强有力和灵活，更多终端用户将考虑把检测任务集成到机器人工作单元。机器人制造商和综合者可以为终端用户提供一些宝贵的建议，使视觉系统的应用不只局限于机器人的引导工作。

"机器人检测系统对零件进行探伤检验，装配和测量工作时，"康耐视公司视觉系统行销主管艾希勒强调说，"视觉系统必须能准确地找到和检测零件。最重要的是，综合者必须保证具有非常高的定位精度且能迅速把信息传回到机器人。"

检测系统要确定零件是否存在。"综合者首先看某个零件是否在装配线上，"机器人视觉技术业务发展经理史蒂夫说，"检测系统可以观察引擎以证实装配完整。例如，在生产线的末端，汽车制造商需要确认滤油器是否安装在引擎上或螺栓是否完全拧紧。"

图 6.2-1　具有视觉的可定位
和分拣任意方位零件的机器人
（FANUC 机器人公司提供）

图 6.2-2　检测工具安装在臂膀上
检测曲轨的机器人
（ABB 公司提供）

机器人"能/不能检测"取决于安装在其胳膊上的照相机，它可移动检测零件上出现的不同特征。

机器人也可用于测量项目。"检测系统可测量组件，但是当测量公差变得越来越小时，检测系统很难满足这些公差，"法鲁克机器人美洲公司智能机器人开发部经理罗尼指出，"机器人的照明设备以及传送给机器人的零件表述就变得更加关键了。当

## Lesson 2　Robotic Inspection

综合者从核实零件的存在转向实际测量时,就增加了检测系统的复杂程度。"

作为机器人检测的一个例子,罗尼说,视觉系统可以确定一个螺母或螺栓是否安装到位,孔是否攻了适当的螺纹。罗尼说,"这些都是机器人检测技术中具有代表性的特征。"

"沙菲进行三维机器人测量需要尺寸公差检测,"沙菲公司总裁沙菲说。"当机器人检测系统用作三坐标测量机时,最终用户不必把零件样本送到实验室进行检测。"沙菲补充说,"机器人检测可以确保在收集零件数据时最终用户能得到正确结果。"

"机器人探伤检测着眼于表面光洁度或确定准确的尺寸,"康耐视公司的艾希勒描述:"最终用户必须界定什么是合格的,什么是不合格的。如果你让流水线上的操作员确定什么是合格的什么是不合格的,他们不可能给出一个相同的答案,检测系统就是要努力找到一个好的解决方案。"

了解检测应用的细节是取得成功的关键。"与零件细节有关的问题和检测要求都是很重要的,"莫托曼公司软件和控制技术负责人卡曼强调指出"综合者必须知道检测要求的详细情况,如需要确定一个极为精细的缺陷或细节,然后选择检测设备,以适应应用。"

同样汽车维修工具公司应用经理史蒂夫·福里斯特注意到,"最终用户必须界定零件的公差范围,并确定适当的方式进行检测。"福雷斯特表示,为了迎接这些挑战,综合者必须使用正确的检测方法。

法兰克公司的埃德·罗尼还强调区分正品与次品的重要性。"最终用户需要准确地了解什么要检验或什么要测量。做机器人检测的工程人员不

图 6.2-3　检测汽车零件的机器人
（**Autotool** 公司提供）

仅需要了解项目处于良好状态时是什么样的,还需要了解项目处于不正常状态时又是什么样子的。当次品出现的时候,该系统还必须妥善处理它。"

罗尼说,建立检测系统时,综合者需要从最终用户获取所有正品和次品的构成正素。"综合者往往要搞一套零件的样本集,但所取的样本集不可能作为产品的长期代表。"罗尼警告。

阿迪尔·沙菲认为,机器人检测系统比传统的检测方案节省成本。"机器人检测提高制造的质量,因为机器人可以检测每一个零件,而不是仅仅检测样本。"传统的质量检验只将百分之一或二的零件送到实验室进行合格检验。"来自质量和成本的压力促进了机器人检测的应用,"沙菲说。

史蒂夫指出检测所有的零件,而不是只检测一个小样本集对 ABB 公司的最终用户来说是非常重要的。"传统的检测是将一些零件送到质量部门,放到一个坐标测量机上进行质量检测。有了机器人检测技术,制造商能够进行在线质量检测。它的优点

是,制造商可以检测每一个零件,而不是仅检测一个百分比。"

能够进行在线检测可以减少废料,节省时间和金钱。丽莎·艾希勒说:"机器人检测技术使得质量检查可以较早地介入到生产过程中,这样最终用户就可以停止在不合格零件上继续增加价值,并且较早地、更为可靠地把它从装配线上撤下来。"艾希勒接着说,除了查找零件外,检测增加了视觉技术应用的复杂性,但利润来源于通过提高质量而节省出来的成本。

虽然检测系统越来越强大且使用方便,但向那些有经验的人学习可以节省时间和金钱。对于那些需要通过视觉系统来提高生产速度的最终用户,找到一个有经验的综合者排在所有待办事项清单的最前面。此外,最终用户还可在检测系统的能力方面多做些实际研究来节省时间和金钱。

"有一个经验丰富的综合者和获得良好的技术支持是很重要的。在一些复杂的应用中,如箱子的随机分拣,最终用户需要有一个对图像处理,编程,高等数学,以及机器人技术都很精通的综合者,"施克公司综合部经理崔京·弗雷迪克森建议。"北美约有1000个视觉系统综合者,只有大约30个经过认证或授权作为施克公司的视觉综合者。"

瓦朗蒂娜机器人公司总裁兼首席执行官安德鲁·瓦朗蒂娜建议,"最终用户第一步要做的是接受良好的培训,获得有关所使用机器人和检测技术本身的知识。如果出现了问题,最终用户必须能够区分是机器人出了问题还是检测系统出了问题。"

ABB公司的史蒂夫·温思特提醒机器人检测系统的最终用户,"用些时间预先确定下应用的要求,此外要避免把检测方案搞得太复杂,并确保最终用户有一个方案来预测,应对和处理有问题的废品。"

阿迪尔·沙菲还强调,必须找到一个经验丰富的视觉综合者。"最终用户应该做好必要的准备工作。那样的话使用检测系统的人就会知道检测系统的运行方式以及预期的解决方案和潜在的结果。这样才能确保检测系统满足用户的需求。"沙菲强调指出,那些第一次购买检测系统的用户应知道他们应用的尺寸公差是多少。

图6.2-4 安装了3维相机进行箱体分拣操作的机器人
(**SICK**公司提供)

"有时演示或测试一下检测系统,可以消除最终用户的忧虑,同时也可以确保系统的正常运行。此外,首次使用的用户应要求参观一家正在使用供应商系统的工厂,"阿迪尔·沙菲建议。"这保证了检测系统的供应商知道他们在做什么,并能让最终用户了解他们会得到什么。"沙菲还提到,检测图像的捕捉时机非常重要,他建议说,在焊接检测应用期间,应该在没有火花的时候进行检测,否则焊接产生的闪光会导致视觉系统在捕捉正在焊接的零件图像时出现问题。

法兰克机器人公司的埃德·罗尼也认为,综合者的经验是至关重要的。"任何经销商都可以出售检测系统,但只有经验丰富的经销商才可能正确的应用它并明白它所包含的问题,"罗尼说。"如果最终用户想让一个综合者来完成这个检测任务,应该先了解一下他在这方面的经验。如果最终用户决定自己做综合检测系统,他们需要做更多的功课并调查一下他们应用的要求是什么。"

莫托曼机器人公司的格雷格·卡曼敦促最终用户"明确界定检测的范围,选择检测设备时,技术指标要比他们要求的指标更高一些。这将为按照需要提高检测精度提供一定的灵活性或克服一些不可预见的困难。"

由于视觉系统变得越来越灵活有效,因此检测系统将能够更好地处理一些由于环境因素造成的问题,比如在机器人操作过程中,由于温度升高会导致其结构尺寸有所变化。阿迪尔·沙菲解释说:"在未来的几年内,检测系统将更加准确,将有更好的温度补偿技术。当机器人温度升高时,一毫米的膨胀就会在检测结果上造成不小的变化。"沙菲也注意到目前市场上有许多三维视觉软件包可供使用。

格雷格·卡曼预计,在视觉和检测设备上取得的进步将进一步简化检测工具和检测步骤。"这种趋势将持续下去,因为随着消费者对优质产品的需求不断增加,每个行业需要检测的数据就更多。"

埃德·罗尼预测基于机器视觉的检测系统将越来越被人所接受。"我看到这样一个趋势,越来越多的人坦然接受应用检测系统。越来越多的公司在机器人导引和纠错技术上使用了视觉系统。他们发现视觉系统比以往任何时候都更具价值。"

如果已经使用了视觉系统为机器人导引,更多的最终用户就会认为该系统可以检测机器人正在做的工作,史蒂夫·韦斯特认为:"更多的人正在考虑如何利用机器人进行检测。如果已经使用了机器人进行物料输送任务,只需另外增加一个传感器就可以使机器人进行检测工作,因此不必增加许多额外的费用。"

施克公司的崔京·弗雷迪克森也看到了检测技术正从二维向三维转换。"今天,施克公司主要使用三角测量技术做三维视觉。""在今后的五年里,随着计算机速度的加快,将会有更多的立体视觉检测技术产生。"电脑速度还是不够快,立体视觉需要更快的处理器在制造所需的周期进行所有的数学运算。提高计算机的处理能力将打开立体视觉检测技术的大门。弗雷迪克森补充说,"今天,我们都知道该怎么获得立体视觉,但发展的瓶颈就是计算机的处理能力。"

在今后几年内,机器人检测将成为主流,这样,大多数制造商能够在日常生产中使用机器人进行检测。此外,还要寻找能同时检测多个项目的视觉系统。

**构词法之合成法**
一、定义:合成法指用两个或更多的词合成一个词。
二、具体构成方式

（一）复合名词

over（[副词]过度）+ exposure（[名词]暴露）= overexposure（[名词]过度暴露）
high（[形容词]高的）+ temperature（[名词]温度）= high-temperature（[名词]高温）

（二）复合形容词

common（[形容词]普通的）+ place（[名词]地区）= commonplace（[形容词]普遍的）
copper（[名词]铜）+ based（[形容词]以…为基础的）= copper-based（[形容词]以铜为基础的）

（三）复合动词

sleep（[动词]睡觉）+ walk（[动词]走路）= sleep-walk（[动词]梦游）

（四）复合代词

her（[代词][宾格]她）+ self（[名词]自己）= herself（[代词]她自己）
it（[代词]它）+ self（[名词]自己）= itself（[代词]她自己）
some（[代词]一些）+ thing（[名词]东西）= something（[代词]某物）

# Unit 7

## Lesson 1  Computer-Aided Design (CAD)

*1*  Computer-aided design (CAD) means the use of a computer to assist in the design of an individual part, such as car, plane and aircraft and so on. The design process often involves computer graphics.

*2*  A CAD system is basically a design tool in which the computer is used to perform various aspects of a design product. The CAD system supports the design process at all stages, conceptual, preliminary, and final design. The designer can test the part in various environmental conditions by using CAD/CAE/CAM system.

*3*  Currently, most CAD systems are using interactive graphics system that is very convenient for beginners. The display of the designed object on a screen is one of the most valuable features of graphics system. The picture of the designed object is usually displayed on the surface of a cathode-ray tube (CRT). However, fortunately, a designer or user do not necessarily know the theory of interactive graphics system. Interactive graphics system has many advantages than traditional graphics system. Interactive graphics system enables the designer or user to study the object by rotating it on the computer screen, separating it into segments, enlarging a specific portion of the object in order to observe it in detail.

*4*  For example, many users like to design the complex part by using Pro/Engineer wildfire edition. When using Pro/E software to design the part, we can perceive the goodness of interactive graphics system. When designed dimensions of part appear error, we can modify it in time by interactive graphics system platform. When designed part need to be observed, we can discern the part on the computer screen by rotating the middle key of mouse.

*5*  The end products of many CAD systems are drawings generated on a plotter interfaced with the computer. One of the most difficult problems in CAD drawings is the elimination of hidden lines. The computer produces the drawing as wire frame diagram. Various methods are used to generate the drawing of the part on the computer screen. One method is to use a geometric modeling approach, in which fundamental shapes and basic elements are used to build the drawing. The lengths and radius of the elements can be modified. For example, a cylinder is a basic element, the subtraction of a cylinder with a specific radius and length will create a hole in the displayed part.

*6* Recently, CAD systems are using the finite element method (FEM). By this approach the object to be analyzed is represented by a model consisting of small elements. The analysis requires the simultaneous solution of many equations. For example, at the present time, finite element analysis (FEA) is the most widely used method in simulating the injection molding process. FEA method has a very important effect on the design and manufacturing process in the plastic injection molding industry in terms of both quality improvement and cost reduction. Therefore, FEA method not only is used to replace real experiments for the sake of cost saving, but also the advantage of FEA method is that the time required to train the artificial neural network is much less than that real experiments because there is no noise existing in the computation data compared with the experimental data. Fig. 7.1 – 1 shows the result of volumetric shrinkage distribution injection molding by FEA method.

Fig. 7.1 – 1 **injection molding by FEA**

*7* The CAD system generates at the design stage a single geometric data base which can be used in all stages of the design and later in the manufacturing, assembly, and so on.

*8* Advanced CAD systems include solid geometry modeling capability, in addition to the wire frame mode diagrams. In recent years, the CAD system technology has improved industry production efficiency. It is a significant step toward the design of the factory of the future.

### ❖ New Words and Phrases

| | |
|---|---|
| assist [ə'sist] v. | 辅助 |
| segment ['segmənt] n. | 部分；段，节 |
| Computer-Aided Design (CAD) | 计算机辅助设计 |
| interactive graphics system | 交互图形系统 |
| traditional graphics system | 非交互图形系统 |
| cathode ['kæθəud] n. | 阴极 |
| ray [rei] n. | 射线 |
| tube ['tju:b] n. | 管，软管 |
| cathode-ray tube (CRT) | 阴极射线管 |
| plotter ['plɔtə] n. | 绘图仪 |
| interfaced ['intə:feist] a. | 界面上的，界面的 |

# Lesson 1   Computer-Aided Design (CAD)

| | |
|---|---|
| interfaced with the computer | 与电脑相连接 |
| frame [freim] n. | 框,结构,骨架 |
| diagram ['daiəgræm] n. | 图解,图表 |
| wire frame diagram | 线架模型 |
| finite element method (FEM) | 有限元分析方法 |
| injection molding process | 注塑成型过程 |
| for the sake of... | 为了…… |
| artificial [ˌɑːtiˈfiʃəl] adj. | 人造的 |
| neural ['njuərəl] adj. | 神经的 |
| artificial neural network | 人工神经网络 |
| volumetric [vɔljuˈmetrik] adj. | 测定体积的 |
| shrinkage ['ʃrinkidʒ] n. | 收缩,缩小,减低 |
| distribution [ˌdistriˈbjuːʃən] n. | 分布 |
| volumetric shrinkage distribution | 体收缩率分布图 |
| geometric [dʒiəˈmetrik] adj. | 几何学的 |
| geometric data base | 几何数据库 |
| solid ['sɔlid] n. | 固体 |
| solid geometry modeling | 实体建模 |
| in addition to | 除此以外 |
| production efficiency | 生产效率 |

## ❖ Notes

1. Currently, most CAD systems are using interactive graphics system that is very convenient for beginners.
   当前,大多数计算机辅助设计系统采用的是交互图形系统方式,对初学者来说非常方便。
2. Interactive graphics system has many advantages than traditional graphics system.
   图形交互系统比传统交互系统有很多优点。
3. FEA method has a very important effect on the design and manufacturing process in the plastic injection molding industry in terms of both quality improvement and cost reduction.
   就制品质量的改进和费用的降低而言,有限元方法对塑料注塑成型的设计与制造有着重要的影响。

**语法补充:both... and... & neither... nor... & either... or... & not... but... &**

**not only... but also...**

一、本语法在注释三中的应用

本注释运用了 both... and... 的搭配,意思是"两者都……",说明了"FEA method"对"quality improvement"和"cost reduction"都有很大影响。

二、对本语法的详细阐述

1. both... and... "既……又……",谓语动词一定要用复数。

   e.g. Both she and I are going to do the cleaning.

   我和她两个都要做清洁工作。

2. neither... nor... "既不……也不……",谓语动词与离它较近的主语保持一致,即就近原则。

   e.g. Neither she nor I am going to do the cleaning.

   我和她都不要做清洁工作。

3. either... or... "或者……或者……",谓语动词与离它较近的主语保持一致,即就近原则。

   e.g. Either she or I am going to do the cleaning.

   要么我要么她要做清洁工作。

4. not... but... "不是……而是……",谓语动词与离它较近的主语保持一致,即就近原则。

   e.g. Not the students, but the teacher is hoping to go there.

   不是这些学生而是这个老师希望去那儿。

5. not only... but also "不但……而且",谓语动词与离它较近的主语保持一致,即就近原则。

   e.g. Not only the students but also their teacher is enjoying the film.

   不仅学生们在欣赏这部影片,他们的老师也在欣赏这部影片。

## *Check your understanding*

Ⅰ. Give brief answers to the following questions.

1. Currently, what system are most CAD systems using?
2. What is one of the most difficult problems in CAD drawings?

Ⅱ. Match the items listed in the following two columns.

diagram          非交互图形系统

tube             计算机辅助设计

Computer-Aided Design   交互图形系统

traditional graphics system   图解,图表

Lesson 1  Computer-Aided Design (CAD)

finite element method　　　　　管,软管
interactive graphics system　　　有限元分析方法
artificial　　　　　　　　　　　 实体建模
distribution　　　　　　　　　　人造的
solid geometry modeling　　　　 分布

## 计算机辅助设计

　　计算机辅助设计就是借助于电子计算机辅助设计一件物品,例如汽车,飞机和飞行器等等。计算机辅助设计常常涉及到计算机图形学。

　　计算机辅助设计系统基本上是一种设计工具,计算机是用来完成设计产品的各个方面。计算机辅助设计系统支持设计过程的每个阶段,包括概念阶段,准备阶段和最终阶段。在不同的工作环境下,设计人员可以通过利用计算机辅助设计、计算机辅助分析和计算机辅助制造系统验证设计产品。

　　当前,大多数计算机辅助设计系统采用的是交互图形系统方式,对初学者来说非常方便。设计的产品显示在屏幕上,是图形交互系统最重要的特征之一。设计产品的模型通常显示在阴极射线管的表面。但是,幸运的是,设计人员和用户并不需要了图形交互系统的工作原理。图形交互系统与传统交互系统相比,有很多优点。图形交互系统使设计人员或用户能够观察设计的产品,方法是在电脑屏幕上旋转它,分割成不同的部分,或者为了详细的观察产品,放大特定的部分。

　　例如,许多用户喜欢利用 Pro/E 软件野火版进行复杂产品的建模。当我们使用 Pro/E 软件野火版进行产品设计时,你可以感觉到图形交互系统的好处。当我们设计的产品尺寸出现错误时,我们可以通过交互图形系统平台来及时修改。当我们需要观察设计的产品时,我们可以通过旋转鼠标的中键来研究它们。

　　许多 CAD 系统的最终产品是与计算机连接的绘图仪中产生的图形。在 CAD 绘图中,最主要的难题之一就是隐藏线的消除。计算机主要是通过线架模型来产生图形。在计算机屏幕上,许多不同的方法被用来产生零件的图形。其中一个方法就是利用几何建模方法来建立零件图形的基本形状和元素。这些元素的长度和半径可以修改。例如,圆柱体是一个基本元素,在显示的零件上减去一个给定半径和长度的圆柱体就可以创建一个孔。

　　最近,许多计算机辅助设计系统都采用了有限元方法。利用有限元方法,分析模型被分解成许多微小单元,分析同时需要解许多方程。例如,在目前,有限元方法是注塑成型模拟过程中最常使用的方法。就制品质量的改进和费用的降低而言,有限元方法对塑料注塑成型的设计与制造有着重要的影响。因此,有限元方法不仅被用来取代

真实的实验以节省费用,而且有限元方法的优势是训练神经网络所需要的时间比真实的实验少,因为与实验数据相比,计算数据没有噪音存在。图 7.1-1 显示了注塑成型有限元分析模型。

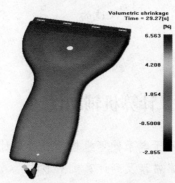

7.1-1　注塑成型有限元分析

在设计阶段,计算机辅助设计系统产生一个几何数据库,这个数据库用于后来的设计,制造和装配等等各个阶段。

先进的计算机辅助设计系统不仅包括线框模型图,还具有实体几何建模能力。最近几年,计算机辅助设计技术已经提高了工业的生产效率,对未来工厂的设计,这是非常重要的一步。

# Lesson 2 Computer aided manufacturing (CAM)

*1* Computer aided manufacturing (CAM) is the use of computer-based software tools that assist engineers and machinists in manufacturing or prototyping product components. CAM is a programming tool that makes it possible to manufacture physical models using computer aided design (CAD) programs. CAM creates real life versions of components designed within a software Package. CAM was first used in 1971 for car body design and tooling.

## Overview

*2* Traditionally, CAM has been considered as a numerical control (NC) programming tool wherein three-dimensional (3D) models of components generated in CAD software are used to generate CNC code to drive numerically controlled machine tools. Although this remains the most common CAM function, CAM functions have expanded to integrate CAM more fully with CAD/CAM/CAE PLM solutions.

*3* As with other "Computer-Aided" technologies, CAM does not eliminate the need for skilled professionals such as Manufacturing Engineers and NC Programmers. CAM, in fact, both leverages the value of the most skilled manufacturing professionals through advanced productivity tools, while building the skills of new professionals through visualization, simulation and optimization tools.

## Early Use of CAM

*4* The first commercial applications of CAM were in large companies in the automotive and aerospace industries, for example, UNISURF in 1971 at Renault (Bezier) for car body design and tooling.

## Historical shortcomings

*5* Historically, CAM software was seen to have several shortcomings that necessitated an overly high level of involvement by skilled CNC machinists. CAM software would output code for the least capable machine, as each machine tool interpreter added on to the standard g-code set for increased flexibility. In some cases, such as improperly set up CAM software or specific tools, the CNC machine required manual editing before the program will run properly. None of these issues were so insurmountable that a thoughtful engineer could not overcome for prototyping or small production runs; G-Code is a simple language. In high production or high precision shops, a different set of problems were encountered where an experienced CNC machinist must both hand-code programs and run CAN software.

6    Integration of CAD with other components of CAD/CAM/CAE PLM environment requires an effective CAD data exchange. Usually it has been necessary to force the CAD operator to export the data in one of the common data formats, such as IGES or STL, that are supported by a wide variety of software. The output from the CAM software is usually a simple text file of G-code, sometimes many thousands of commands long, that is then transferred to a machine tool using a direct numerical control (DNC) program.

7    CAM packages could not, and still cannot, reason as a machinist can. They could not optimize tool paths to the extent required of mass production. Users would select the type of tool, machining process and paths to be used. While an engineer may have a working knowledge of g-code programming, small optimization and wear issues compound over time. Mass-produced items that require machining are often initially created through casting or some other non-machine method. This enables hand-written, short, and highly optimized g-code that could not be produced in a CAM package.

8    At least in the United States, there is a shortage of young, skilled machinists entering the workforce able to perform at the extremes of manufacturing; high precision and mass production. As CAM software and machines become more complicated, the skills required of a machinist advance to approach that of a computer programmer and engineer rather than eliminating the CNC machinist from the workforce.

**Typical areas of concern:**

9    1. High speed machining, including streamlining of tool paths.
10   2. Multi-function machining.
11   3. 5 Axis machining.
12   4. Ease of use.

**Overcoming historical shortcomings**

13   Over time, the historical shortcomings of CAM are being attenuated, both by providers of niche solutions and by providers of high-end solutions. This is occurring primarily in three arenas:

14   1. Ease of use.
15   2. Manufacturing complexity.

*16*   3. Integration with PLM and the extended enterprise.

**Ease in use**

*17*   For the user who is just getting started as a CAM user, out-of-the-box capabilities providing Process Wizards, templates, libraries, machine tool kits, automated feature based machining and job function specific tailorable user interfaces build user confidence and speed the learning curve. User confidence is further built on 3D visualization through a closer integration with the 3D CAD environment, including error-avoiding simulations and optimizations.

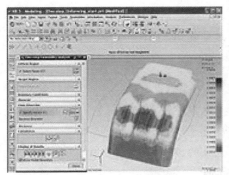 

Fig. 7.2 – 1   **Formability analysis example**      Fig. 7.2 – 2   **Mold and tooling example**

**Manufacturing complexity**

*18*   The manufacturing environment is increasingly complex. The need for CAM and PLM tools by the manufacturing engineer, NC programmer or machinist is similar to the need for computer assistance by the pilot of modern aircraft systems. The modern machinery cannot be properly used without this assistance. Today's CAM systems support the full range of machine tools including: turning, 5 axis machining and wire EDM. Today's CAM user can easily generate streamlined tool paths, optimized tool axis tilt for higher feed rates and optimized Z axis depth cuts as well as driving non-cutting operations such as the specification of probing motions.

**Integration with PLM and the extended enterprise**

*19*   Today's competitive and successful companies have used PLM to integrate manufacturing with enterprise operations from concept through field support of the finished product. To ensure ease of use appropriate to user objectives, modern CAM solutions are scalable from a stand-alone CAM system to a fully integrated multi CAD 3D solution-set. These solutions are created to meet the full needs of manufacturing personnel including part planning, shop documentation, resource management and data management and exchange.

## ❖ New Words and Phrases

| | |
|---|---|
| computer aided manufacturing | 计算机辅助制造 |
| machinist [məˈʃiːnist] n. | 机械师 |
| prototype [ˈprəutətaip] n. | 原型 |
| numerical control | 数字控制 |
| dimensional [diˈmenʃənəl] adj. | 空间的 |
| three-dimensional | 三维 |
| visualization [ˌvizjuəlaiˈzeiʃən] n. | 可视化 |
| visualization tool | 可视化工具 |
| simulation [ˌsimjuˈleiʃən] n. | 模拟，仿真 |
| optimization [ˌɔptimaiˈzeiʃən] n. | 最佳化，优化 |
| necessitate [niˈsesiteit] v. | 使……成为必需，需要 |
| flexibility [ˌfleksəˈbiliti] n. | 柔韧性 |
| insurmountable [ˌinsəˈmauntəbl] adj. | 不能克服的，难以对付的 |
| increased flexibility | 增加柔性 |
| precision [priˈsiʒən] n. | 精度 |
| integration [ˌintiˈgreiʃən] n. | 集成 |
| template [ˈtemplit] n. | 模板，样板 |
| interface [ˈintə(ː)ˌfeis] n. | 界面，接触面 |
| tailorable [ˈteilərəbl] adj. | (衣料等)可裁制成衣的 |
| streamlined [ˈstriːmlaind] adj. | 流线型的 |
| CAM package | 计算机辅助制造包 |
| 3D visualization | 三维可视化 |
| IGES or STL | 文件格式 |
| direct numerical control (DNC) program | 直接数字控制程序 |
| at least | 至少 |
| Wire EDM | 线切割加工 |
| data management and exchange | 数据管理和交换 |

## ❖ Notes

1. Computer aided manufacturing (CAM) is the use of computer-based software tools that assist engineers and machinists in manufacturing or prototyping product components.

# Lesson 2　Computer aided manufacturing (CAM)

计算机辅助制造是以计算机软件作为工具,来协助工程师和机械工制造产品部件或其原型。

2. As with other "Computer-Aided" technologies, CAM does not eliminate the need for skilled professionals such as Manufacturing Engineers and NC Programmers。

与其他的计算机辅助技术一样,计算机辅助制造也没有减少熟练的制造工程师和数控编程员等专业人员的需求。

**语法补充：as 的用法**

一、本语法在注释二中的应用

注释二中"As with other 'Computer-Aided" technologies'运用 as 引导了一个定语从句,告诉读者下文所述的 CAM 的特征与其他计算机辅助技术是一样的。

二、对本语法的详细阐述

1) as 表示"当……的时候",随着……,一边……一边……,引导时间状语从句。

e.g. I thought of it just as you opened your mouth.

你刚张嘴,我就想到这一点了。

e.g. As time went by, she became more and more worried.

随着时间的流逝,她越来越着急。

e.g. The girl dances as she sings on the stage.

女孩在舞台上边唱歌边跳舞。

2) as = since, 作"既然"、"由于"解,引导原因状语从句,如:

e.g. As / Since you're not feeling well, you may stay at home.

你既然不舒服,就待在家里吧。

3) as = in the way that, 作"按照……的方式"解,引导方式状语从句。

e.g. Remember, you must do everything as I do.

记住,你做任何事都必须按照我做事的方式。

4) as 作关系代词,引导定语从句,作"正如"解。

e.g. As is well-known, Taiwan belongs to China.

正如大家所知道的,台湾属于中国。

5) as 作"作为"解

e.g. As a leader, I'll take the lead in everything.

作为领袖,任何事情都由我来领头。

3. In some cases, such as improperly set up CAM software or specific tools, the CNC machine required manual editing before the program will run properly.

在某些情况下,如计算机辅助软件或专门工具设立不当,程序必须经过人工编辑,数控机床才能正确运行。

4. Integration of CAD with other components of CAD/CAM/CAE PLM environment requires
an effective CAD data exchange.

集成计算机辅助设计与其他的计算机辅助设计、计算机辅助制造和计算机辅助分析组件的产品生命周期管理环境需要一个有效的计算机辅助设计数据交换技术。

## *Check your understanding*

Ⅰ. Give brief answers to the following questions.
  1. What is CAM?

Ⅱ. Match the items listed in the following two columns.

| | |
|---|---|
| machinist | 数字控制 |
| CAM packages | 集成 |
| interface | 机械师 |
| numerical control | 精度 |
| visualization tools | 界面 |
| integration | 柔韧性 |
| flexibility | 可视化工具 |
| precision | 计算机辅助制造 |
| computer aided manufacturing | 计算机辅助制造包 |

# 计算机辅助制造（CAM）

计算机辅助制造(CAM)是以计算机软件作为工具,来协助工程师和机械工制造产品部件或其原型。计算机辅助制造是一种编程工具,通过使用计算机辅助设计程序来制造物理原型。通过使用软件包,计算机辅助制造创造了设计零件的立体模型。在1971年,计算机辅助制造第一次用于汽车体设计与加工。

## 概述

传统上,计算机辅助制造被认为是一种数字控制编程工具,通过利用计算机辅助设计软件来设计三维零件模型,产生数字控制代码,来驱动数控机床。尽管这仍是计算机辅助制造最普通的功能,但是通过充分利用计算机辅助设计、计算机辅助制造、计算机辅助分析和产品生命周期管理解决方案,计算机辅助制造的功能已经扩展到集成制造方面。

与其他的计算机辅助技术一样,计算机辅助制造并没有减少熟练的制造工程师和数控编程员等专业人员的需求。事实上,计算机辅助制造是通过先进的生产工具来体现最熟练的制造专业人员的价值,同时通过可视化、模拟和优化工具,来培养新技术人员的技能。

### 早期使用的计算机辅助制造

第一个商业应用的计算机辅助制造是在从事汽车和航空航天行业的大公司,例如,UNISURF系统于1971年应用在雷诺的车身设计和加工上。

### 原始缺点

从历史上看,计算机辅助制造软件有好几个缺点,这使得熟练数控机械师的深度参与成为必要。计算机辅助制造软件为功能单一的机床输出代码,同时为了增加柔性,每台机床在标准G代码中都添加了解释程序。在某些情况下,如计算机辅助制造软件或专门工具设立不当时,程序必须经过人工编辑,才能正确运行。这些问题并非难以克服,肯动脑筋的工程师在进行原型设计或批量生产能够克服这些困难。G代码是一个简单的语言。在高精度和高产量工厂,一位经验丰富的数控机械师遇到一组不同的问题,都必须通过手工编写代码和同步运行软件来解决。

集成计算机辅助设计与其他的计算机辅助设计、计算机辅助制造和计算机辅助分析组件的产品生命周期管理环境需要一个有效的计算机辅助设计数据交换技术。通常有必要强制规定计算机辅助设计运营商使用一个共同的数据格式导出数据,如IGES或STL,这些格式大多数软件都支持它们。计算机辅助制造软件的输出通常是一个简单的G代码的文件格式,有时有成千上万条指令,利用直接数字控制程序将这些指令转移到机床上。

计算机辅助设计制造软件包从前不可能,现在也不能像机械师那样进行思维。它们不能优化刀具路径来达到大批量生产的需求。用户可以选择工具类型、加工过程工艺和加工轨迹。虽然工程师可能具有G代码编程的运作知识,随着时间的推移,小的优化和磨合问题将伴随着进行。通常是利用铸造或非机械加工方法来制造需要机械加工的一些大批量生产项目。这使得计算机辅助设计软件包不能够产生的手写的、短的和最佳的G代码成为可能。

至少在美国,进入工场能够进行极端制造、高精度和大批量生产的年轻且熟练的机械师还是很短缺的。因为计算机辅助制造软件越来越复杂,对机械师的技能要求接近于计算机程序编程人员和工程师,而不是减少数控机械师的数量。

### 典型的关注范围:

1. 高速加工,包括流线型的刀具轨迹

2. 复合功能加工
3. 5 轴加工
4. 使用的简便性

**克服原始缺点**

随着时间的推移,计算机辅助制造软件的原始缺点正在削弱,作出贡献者包括提供小型解决方案者,以及提供高端解决方案者。这主要发生在三个主要领域:
1. 易用性
2. 制造复杂性
3. 集成的产品生命周期管理和扩展型企业

**易用性**

为了计算机辅助制造软件的初始用户者,边界盒尽可能地提供加工向导、模板、库、机床工具、基于特征的自动化加工。用户界面的工作功能是建立用户对我们的信任和快速的学习曲线图表。通过更紧密的集成三维计算机辅助设计环境,包括避错模拟和优化,用户对我们的信任,进一步建立在三维可视化上。

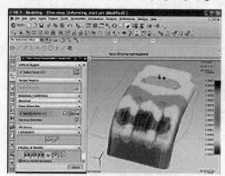

图 7.2-2 成形分析实例    图 7.2-3 模具和工具实例

**制造复杂性**

制造环境是越来越复杂。制造工程师、数控编程员和机械师需要计算机辅助制造和产品生命周期管理工具类似于现代航空体系下的飞行员需要计算机协作。没有这种协作,现代机器不能正常使用。今天的计算机辅助制造系统支持全系列的数控机床,其中包括,车削、5 轴加工和线切割机。今天的计算机辅助制造用户可以轻松的生成精简的刀具轨迹,优化工具轴倾斜角度以获得更高的进给速度,需要优化 Z 轴切削深度,操纵无切割运转,如规范的探测动作。

Lesson 2  Computer aided manufacturing (CAM)　　　　　　　　　　　　　　　　　　*179*

**集成的产品生命周期管理和扩展型企业**

　　当今,有竞争力的和成功的公司都已经使用了产品生命周期管理,从原始概念直至最终产品的生产支持,全面集成加工制造和企业经营。为了确保容易达到用户的目标,现代化的计算机辅助制造解决方案可从一个独立的计算机辅助制造系统扩展到一个完全集成的计算机辅助设计三维的解决方案集。这些解决方案都是为了充分满足生产人员的需要,包括部件规划,文件资源、资源管理和数据管理和交换。

## Lesson 3　Computer Numerical Control

*1*　Most CNC milling machines (also called machining centers) are computer controlled vertical mills with the ability to move the spindle vertically along the Z-axis. This extra degree of freedom permits their use in engraving applications, and 2.5D surfaces such as relief sculptures. When combined with the use of conical tools or a ball nose cutter, it also significantly improves milling precision without impacting speed, providing a cost-efficient alternative to most flat-surface hand-engraving work.

*2*　CNC machines can exist in virtually any of the forms of manual machinery, like horizontal mills. The most advanced CNC milling-machines, the 5-axis machines, add two more axes in addition to the three normal axes (XYZ). Horizontal milling machines also have a C or Q axis, allowing the horizontally mounted workpiece to be rotated, essentially allowing asymmetric and eccentric turning. The fifth axis (B axis) controls the tilt of the tool itself. When all of these axes are used in conjunction with each other, extremely complicated geometries, even organic geometries such as a human head can be made with relative ease with these machines. But the skill to program such geometries is beyond that of most operators. Therefore, 5-axis milling machines are practically always programmed with CAM.

Fig. 7.3 – 1　**Five-axis machining center with rotating table and computer interface**

*3*　With the declining price of computers, free operating systems such as Linux, and open source CNC software, the entry price of CNC machines has plummeted. For example, Sherline, Prazi, and others make desktop CNC milling machines that are affordable by hobbyists.

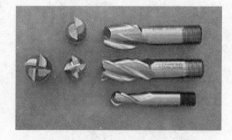

Fig. 7.3 – 2　**Endmills used for cutting operations in a milling machine (high speed steel with cobalt)**

*4*　There is some degree of standardization of the tooling used with CNC Milling Machines and to a much lesser degree with manual milling machines.

5   CNC Milling machines will nearly always use SK (or ISO), CAT, BT or HSK tooling. SK tooling is the most common in Europe, while CAT tooling, sometimes called V-Flange Tooling, is the oldest variation and is probably still the most common in the USA. CAT tooling was invented by Caterpillar Inc. of Peoria, Illinois in order to standardize the tooling used on their machinery. CAT tooling comes in a range of sizes designated as CAT-30, CAT-40, CAT-50, etc. The number refers to the Association for Manufacturing Technology (formerly the National Machine Tool Builders Association (NMTB) Taper size of the tool.

6   An improvement on CAT Tooling is BT Tooling, which looks very similar and can easily be confused with CAT tooling. Like CAT Tooling, BT Tooling comes in a range of sizes and uses the same NMTB body taper. However, BT tooling is symmetrical about the spindle axis, which CAT tooling is not. This gives BT tooling greater stability and balance at high speeds. One other subtle difference between these two toolholders is the thread used to hold the pull stud. CAT Tooling is all Imperial thread and BT Tooling is all Metric thread. Note that this affects the pull stud only, it does not affect the tool that they can hold, both types of tooling are sold to accept both imperial and metric sized tools.

Fig. 7.3 - 3   **CAT - 40 Toolholder**

7   SK and HSK tooling, sometimes called "Hollow Shank Tooling", is much more common in Europe where it was invented than it is in the United States. It is claimed that HSK tooling is even better than BT Tooling at high speeds. The holding mechanism for HSK tooling is placed within the (hollow) body of the tool and, as spindle speed increases, it expands, gripping the tool more tightly with increasing spindle speed. There is no pull stud with this type of tooling.

8   The situation is quite different for manual milling machines — there is little standardization. Newer and larger manual machines usually use NMTB tooling. This tooling is somewhat similar to CAT tooling but requires a drawbar within the milling machine. Furthermore, there are a number of variations with NMTB tooling that make interchangeability troublesome.

9   Two other tool holding systems for manual machines are worthy of note: They are the R8 collet and the Morse Taper #2 collet. Bridgeport Machines of Bridgeport, Connecticut so dominated the milling machine market for such a long time that their machine "The

Bridgeport" is virtually synonymous with "Manual milling machine." The bulk of the machines that Bridgeport made from about 1965 onward used an R8 collet system. Prior to that, the bulk of the machines used a Morse Taper #2 collet system.

10　As an historical footnote: Bridgeport is now owned by Hardinge Brothers of Elmira, New York.

Fig. 7.3 - 4　**Boring head on Morse Taper Shank**

## ❖ New Words and Phrases

numerical [nju(ː)'merikəl] adj.　　　数字的
spindle ['spindl] n.　　　轴
axis ['æksis] n.　　　轴
sculpture ['skʌlptʃə] n.　　　雕塑
combine with　　　把……结合起来
conical ['kɔnikəl] adj.　　　圆锥体的
precision [pri'siʒən] n.　　　精确,精密度
engrave [in'greiv] v.　　　雕刻
hand-engraving　　　手工雕刻
workpiece ['wəːkpiːs] n.　　　工件
interface ['intə(ː)feis] n.　　　界面,接触面
asymmetric [ˌeisi'metrik] adj.　　　不对称的
tilt [tilt] n.　　　倾斜
conjunction [kən'dʒʌŋkʃən] n.　　　连接,连合
geometry [dʒi'ɔmitri] n.　　　几何(学)
decline [di'klain] v.　　　降低
plummet ['plʌmit] vi.　　　垂直落下
desktop ['desktɔp] n.　　　[计算机]桌面
toolholder ['tuːlˌhəuldə(r)] n.　　　刀柄
shank [ʃæŋk] n.　　　胫
standardization [ˌstændədai'zeiʃən] n.　　　标准化
designate ['dezigneit] v.　　　指定,标示
drawbar ['drɔːbaː] n.　　　挂钩,拉杆
variation [ˌvɛəri'eiʃən] n.　　　变化

## Lesson 3  Computer Numerical Control

interchangeability [ˌintə(ː)tʃeindʒəˈbiliti] n.　　可交换性
troublesome [ˈtrʌblsəm] adj.　　令人烦恼的,讨厌的
collet [ˈkɔlit] n.　　筒夹,夹头
dominate [ˈdɔmineit] v.　　支配,占优势
virtually [ˈvɜːtjuəli] adv.　　几乎,实际上
synonymous [siˈnɔniməs] adj.　　同义的
onward [ˈɔnwəd] adj.　　向前的,前进的 adv. 向前,前进,在先
footnote [ˈfutnəut] n.　　脚注(给……作脚注)

## ❖ Notes

1. When combined with the use of conical tools or a ball nose cutter, it also significantly improves milling precision without impacting speed, providing a cost-efficient alternative to most flat-surface hand-engraving work.
当与圆锥刀具或球头刀结合使用时,它在没有影响速度的基础上,显著地提高了铣削精度。为平面加工提供了一种划算的选择。

**语法补充:when 引导时间状语从句时的省略**
一、本语法在注释一中的应用
　　"When combined with the use of conical tools or a ball nose cutter"是一个由关系副词 when 引导的时间状语从句,when 后面省略了"it is",省略的原因是这个时间状语从句的主语和紧接其后的主句的主语一致,省略之后不会引起歧义,且使句子更加简洁。
二、对本语法的详细阐述
　　时间状语从句的其他引导词也可以如此省略。
e. g. When (he was) very young, he began to learn to play the guitar. (when 引导时间状语从句)
　　他很小的时候,就开始学习弹吉他。
e. g. While (I was) at college, I began to know him, a strange but able student. (while 引导时间状语从句)
　　我在上大学时就开始认识他,一个奇怪但有能力的学生。
e. g. Don't come in until (you are) asked to. (until 引导时间状语从句)
　　不叫你请你不要进来。
e. g. Whenever (it is) possible, you should come and help. (whenever 引导时间状语从句)
　　不管什么时候只要有可能你就该来帮忙。
e. g. You should let us know the result as soon as (it is) possible. (as soon as 引导时间状

语从句)

你应该尽快让我们知道结果。

2. Horizontal milling machines also have a C or Q axis, allowing the horizontally mounted workpiece to be rotated, essentially allowing asymmetric and eccentric turning.

卧式机床也有 C 或 Q 轴,允许水平安装的工件进行旋转运动,尤其适合不对称和不规则车削加工。

3. Furthermore, there are a number of variations with NMTB tooling that make interchangeability troublesome.

另外,NMTB 刀具还有一些变化,使其互换较困难。

## Check your understanding

Ⅰ. Give brief answers to the following questions.
   1. What are the most advanced CNC milling-machines?
   2. What tooling will CNC Milling machines use?

Ⅱ. Match the items listed in the following two columns.

| | |
|---|---|
| drawbar | 雕刻 |
| engrave | 挂钩,拉杆 |
| workpiece | 轴 |
| spindle | 刀柄 |
| designate | 工件 |
| toolholder | 筒夹,夹头 |
| collet | 指定,标示 |
| shank | 可交换性 |
| interchangeability | 胫 |

## CNC(计算机数字控制)

大多数的数控加工中心(也叫机械加工中心)是由计算机控制的拥有能沿 Z 轴垂直移动能力的立式铣床。这样另外的自由度允许它们在雕刻、2.5D 表面加工中(如浮雕)的应用。当与圆锥刀具或球头刀结合使用时,它在没有影响速度的基础上,显著地提高了铣削精度。为平面加工提供了一种划算的选择。

# Lesson 3  Computer Numerical Control

图 7.3-1  5 轴联动加工中心拥有可旋转的台面和计算机界面

数控加工中心事实上可以任何机械形式存在,像卧式加工中心。5 轴联动就是最先进的数控加工中心,它在普通的 XYZ 坐标机床上多了 2 个轴。卧式机床也有 C 或 Q 轴,允许水平安装的工件进行旋转运动,尤其适合不对称和不规则的车削加工。第五个轴(B 轴)控制刀具自身的倾斜,当这些轴相互联动时,应用这些机器加工一些非常复杂的几何体,甚至是如人头一般的组织形体,都是非常容易的。但是对这些几何形体进行编程的技能已经超出了大多数操作者的能力范围,因此 5 轴联动加工中心的编程实际上总是借助于计算机辅助制造(CAM)技术。

随着计算机价格的回落,免费操作系统的提供,如 Linux,开源 CNC 软件,数控机床的报价也随之下降。例如 Sherline,Prazi 和其他一些台式数控机床的价格让业余爱好者们都能承担。

数控机床使用的刀具具有一定的标准,而手工机床刀具不是那么严格。

数控机床常使用 SK (or ISO), CAT, BT or HSK 几种刀具。SK 在欧洲最常见,CAT 通常被称作 V 型法兰刀具,它是历史最悠久的一种类型,也几乎是美国最通用的刀具。CAT 刀具是由伊丽诺斯州皮里亚卡特皮勒公司发明,为了规范应用在它们机器上的刀具。CAT 刀具有

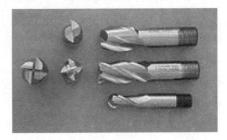

图 7.3-2  在铣削加工中使用的钴高速钢端铣刀

一系列尺寸,包括 CAT-30,CAT-40,CAT-50 等等。这些数据参考了制造技术协会(原国家机械刀具制造协会)刀具的锥度尺寸规格。

一种 CAT 刀具的升级版本就是 BT 刀具,看起来和 CAT 很相似,而且很容易与其混淆。与 CAT 一样,BT 拥有各种尺寸且使用同样的 NMTB 标准刀柄。但 BT 刀具是相对与轴线对称的,而 CAT 刀具并不这样。这让 BT 刀在高速状态下保持稳定与平衡。另外一个微小区别是这两种刀柄连接拉钉的螺纹不一样。CAT 用的是英制螺纹,而 BT 则是公制螺纹。需要指出的是这只对拉钉有影响,却不影响它们能够夹持的刀具。出售的这两种刀具系统都可以装夹英制和公制尺寸的刀具。

SK 和 HSK 刀具被称作挖孔刀,相比于美国,在它的故乡欧洲使用更广泛些。HSK 在高速切削上比 BT 刀更好,HSK 的夹紧机构放置在孔中央,随着轴转速增加,它随之扩大,牢牢地抓紧刀具,这种刀具里没有拉杆。

手工机床的情形完全不同——几乎没有标准。越新越大的手工机器一般都采用 NMTB 标准刀具。这种刀具和 CAT 有点类似,但在铣床内需要一个拉杆。另外,NMTB 刀具还有一些变化使得互换较困难。

图 7.3-3 CAT-40 的刀柄

还有两种用于手工机器的夹具值得一提,R8 夹头和莫氏#2 夹头。康湿狄格州桥港市桥港机械公司在很长一段时间内控制着铣床市场,以致桥式几乎成了手工机床的同义词。在 1965 年之后该公司生产的机器主要的是 R8 系统,而在此之前,大部分机器的用的莫氏 #2 系统。

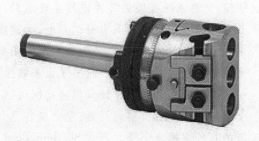

图 7.3-4 镗孔用的莫氏刀杆

历史记载:Bridgeport(桥式铣床)现在由纽约州 Elmira 市,New York. 的哈叮兄弟所拥有。

## 一、构词法之混合法
(一)定义:合成法指将两个词混合或各取一部分紧缩而成一个新词,前半部分表属性,后半部分表主体。
(二)具体构成方式
biology([名词]生物)& material([名词]材料)→ biomaterial([名词]生物材料)
electron([名词]电子)& slag([名词]熔渣)→ electroslag([名词]电炉渣)
electron([名词]电子)& plating([名词]镀层)→ electroplating([名词]电镀术)

## 二、构词法之首字母缩略法
(一)定义:首字母缩略法指取词组中每个词的第一个字母代表这个词组构成新词。
(二)具体构成方式
GTAW – Gas tungsten arc welding([名词]钨极气体保护电弧焊)
FCAW - flux-cored arc welding([名词]药芯焊丝电弧焊)

# Unit 8

## Lesson 1  3D Printing Technology

*1*  3D printing, also known as additive manufacturing (AM), refers to various processes used to synthesize a three-dimensional object. In 3D printing, successive layers of material are formed under computer control to create an object. These objects can be of almost any shape or geometry, and are produced from a 3D model or other electronic data source. A 3D printer is a type of industrial robot.

*2*  Futurologists such as Jeremy Rifkin [3] believe that 3D printing signals the beginning of a third industrial revolution, [4] succeeding the production line assembly that dominated manufacturing starting in the late 19th century. Using the power of the Internet, it may eventually be possible to send a blueprint of any product to any place in the world to be replicated by a 3D printer with "elemental inks" capable of being combined into any material substance of any desired form.

Fig. 8.1 – 1
**Additive Manufacturing**

*3*  3D printing in the term's original sense refers to processes that sequentially deposit material onto a powder bed with inkjet printer heads. More recently, the meaning of the term has expanded to encompass a wider variety of techniques such as extrusion and sintering-based processes. Technical standards generally use the term additive manufacturing for this broader sense.

### 1. General principles

*4*  3D printable models may be created with a computer aided design (CAD) package, via a 3D scanner or by a plain digital camera and photogrammetry software. 3D printed models created with CAD results in reduced errors and can be corrected before printing, allowing verification in the design of the object before it is printed.

*5*  The manual modeling process of preparing geometric data for 3D computer graphics is similar to plastic arts such as sculpting. 3D scanning is a process of collecting digital data on the shape and appearance of a real object, creating a digital model based on it.

*6*  Before printing a 3D model from an STL file, it must first be examined for "manifold errors". This step is usually called the "fixup." Generally STLs that have been produced from a model obtained through 3D scanning often have many manifold errors in them that need to be rectified. Examples of these errors are surfaces that do not connect, or gaps in the models.

7   Once completed, the STL file needs to be processed by a piece of software called a "slicer," which converts the model into a series of thin layers and produces a G-code file containing instructions tailored to a specific type of 3D printer (FDM printers). This G-code file can then be printed with 3D printing client software which loads the G-code, and uses it to instruct the 3D printer during the 3D printing process.

8   Printer resolution describes layer thickness and X-Y resolution in dots per inch (dpi) or micrometers (μm). Typical layer thickness is around 100 μm (250 DPI), although some machines can print layers as thin as 16 μm (1,600 DPI). X-Y resolution is comparable to that of laser printers. The particles (3D dots) are around 50 to 100 μm (510 to 250 DPI) in diameter.

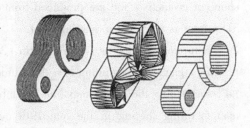

Fig. 8.1-2   The Solid Models (a) Solid Model, (b) an STL File, and (c) Sliced Layers.

9   Construction of a model with contemporary methods can take anywhere from several hours to several days, depending on the method used and the size and complexity of the model. Additive systems can typically reduce this time to a few hours, although it varies widely depending on the type of machine used and the size and number of models being produced simultaneously.

10   Traditional techniques like injection moulding can be less expensive for manufacturing polymer products in high quantities, but additive manufacturing can be faster, more flexible and less expensive when producing relatively small quantities of parts. 3D printers give designers and concept development teams the ability to produce parts and concept models using a desktop size printer.

11   The printer-produced resolution is sufficient for many applications, for example, printing a slightly oversized version of the desired object in standard resolution, then removing material with a higher-resolution subtractive process to achieve greater precision. Some printable polymers allow the surface finish to be smoothed and improved using chemical vapor processes.

12   Some additive manufacturing techniques are capable of using multiple materials in the course of constructing parts. These techniques are able to print in multiple colors and color

# Lesson 1  3D Printing Technology

combinations simultaneously, and would not necessarily require painting.

*13*  Some printing techniques require internal supports to be built for overhanging features during construction. These supports must be mechanically removed or dissolved upon completion of the print.

## 2. 3D printing process

*14*  Several 3D printing processes have been invented since the late 1970s. The printers were originally large, expensive, and highly limited in what they could produce.

*15*  One of 3D printing approaches is the selective fusing of materials in a granular bed. The technique fuses parts of the layer and then moves upward in the working area, adding another layer of granules and repeating the process until the workpiece has been built up. This process uses the unfused media to support overhangs and thin walls in the part being produced, which reduces the need for temporary auxiliary supports for the piece. A laser is typically used to sinter the media into a solid. Examples include selective laser sintering (SLS), with both metals and polymers, and direct metal laser sintering (DMLS).

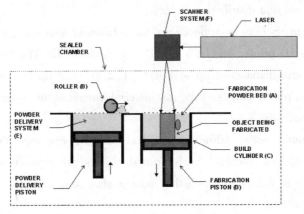

Fig. 8.1-3  **Selective Laser Sintering**

*16*  Selective laser melting (SLM) does not use sintering for the fusion of powder granules but will completely melt the powder using a high-energy laser to create fully dense materials in a layer-wise method that has mechanical properties similar to those of conventional manufactured metals.

*17*  Another method consists of an inkjet 3D printing system. The printer creates the model one layer at a time by spreading a layer of powder (plaster, or resins) and printing a binder in the cross-section of the part using an inkjet-like process. This is repeated until every layer has been printed. This technology allows the printing of full color prototypes, overhangs, and elastomer parts. The strength of bonded powder prints can be enhanced with wax or thermoset polymer impregnation.

## 3. 3D Printers

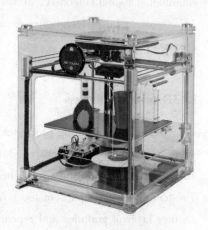

Fig. 8.1-4   3DTouch™
3D Printer-Double Head

*18*   As in October 2012, additive manufacturing systems were on the market that ranged from $2,000 to $500,000 in price and were employed in industries including aerospace, architecture, automotive, defense, and medical replacements, among many others. For example, General Electric uses the high-end model to build parts for turbines. Many of these systems are used for rapid prototyping, before mass production methods are employed.

*19*   As the costs of 3D printers have come down they are becoming more appealing financially to use for self-manufacturing of personal products. In addition, 3D printing products at home may reduce the environmental impacts by reducing material use and distribution impacts.

*20*   Large 3D printers have been developed for industrial, education, and business uses. A large delta-style 3D printer was built in 2014 by SeeMeCNC. The printer is capable of making an object with diameter of up to 4 feet (1.2 m) and up to 10 feet (3.0 m) in height. It also uses plastic pellets as the raw material instead of the typical plastic filaments used in other 3D printers.

*21*   Microelectronic device fabrication methods can be employed to perform the 3D printing of nanoscale-size objects. Such printed objects are typically grown on a solid substrate, e.g. silicon wafer, to which they adhere after printing as they're too small and fragile to be manipulated.

## ❖ New Words and Phrases

| | | |
|---|---|---|
| additive manufacturing | | 增材制造 |
| synthesize['sɪnθəsaɪz] | v. | 合成;综合 |
| signal['sɪgnəl] | v. | 表示 |
| replicate['replɪkeɪt] | v. | 复制 |
| elemental ink | | 元素墨水 |
| inkjet printer['ɪŋkdʒetprɪntər] | n. | 喷墨式打印机 |
| extrusion [ɪk'stru:ʒn] | n. | 挤出 |
| sinter['sɪntə] | v. | 烧结 |

## Lesson 1  3D Printing Technology

| | |
|---|---|
| photogrammetry [ˌfoʊtəˈgræmɪtriː] n. | 照相测量法 |
| geometric [ˌdʒiːəˈmetrɪk] adj. | 几何学的 |
| fixup [ˈfɪksʌp] n. | 修补 |
| rectify [ˈrektɪfaɪ] v. | 改正 |
| slicer [ˈslaɪsə] n. | 切片 |
| tailor [ˈteɪlər] v. | 专门定制 |
| resolution [ˌrezəˈluːʃn] n. | 分辨率 |
| flexible [ˈfleksəbl] adj. | 灵活的 |
| subtractive [səbˈtræktɪv] adj. | 减去的 |
| subtractive process | 减材过程 |
| surface finish | 表面光洁度 |
| vapor [ˈveɪpə] n. | 蒸汽 |
| chemical vapor process | 化学气相过程 |
| painting [ˈpeɪntɪŋ] n. | 涂漆, 着色 |
| dissolve [dɪˈzɒlv] v. | 使溶解 |
| granular [ˈgrænjələ(r)] adj. | 由小粒而成的;粒状的 |
| auxiliary [ɔːgˈzɪliəri] adj. | 辅助的 |
| selective laser sintering | 选择性激光烧结 |
| direct metal laser sintering | 直接针对金属的激光烧结 |
| dense [dens] adj. | 密集的;稠密的 |
| binder [ˈbaɪndə(r)] n. | 粘合剂 |
| cross-section [ˈkrɒsˈsekʃən] n. | 横断面 |
| prototype [ˈprəʊtətaɪp] n. | 原型 |
| overhang [ˈəʊvəhæŋ] n. | 突出部分;悬垂部分 |
| elastomer [ɪˈlæstəmə(r)] n. | 弹性体 |
| impregnation [ˌɪmpregˈneɪʃn] n. | 注入;浸渍 |
| turbine [ˈtɜːbaɪn] n. | 涡轮 |
| diameter [daɪˈæmɪtə(r)] n. | 直径 |
| pellet [ˈpelɪt] n. | 颗粒状物 |
| filament [ˈfɪləmənt] n. | 细丝;细线 |
| fabrication [ˌfæbrɪˈkeɪʃn] n. | 制作 |
| substrate [ˈsʌbstreɪt] n. | 基底 |
| wafer [ˈweɪfə(r)] n. | 晶片 |
| adhere [ədˈhɪə(r)] v. | 粘附 |

## ❖ Notes

1. Using the power of the Internet, it may eventually be possible to send a blueprint of any product to any place in the world to be replicated by a 3D printer with "elemental inks" capable of being combined into any material substance of any desired form.

   借助于互联网的功能,它有可能最终实现如下场景,即将任何一个产品的设计图发送到全世界的任何一个地方,都能使用一台 3D 打印机将产品复制出来。打印机配有"元素墨水",有能力将其组合成任意所需形式的任何物体。

2. Once completed, the STL file needs to be processed by a piece of software called a "slicer," which converts the model into a series of thin layers and produces a G-code file containing instructions tailored to a specific type of 3D printer (FDM printers).

   一旦完成,STL 文件需要由一款被称为"切片"的软件进行后处理。这种软件能够把模型转化为一系列的薄片层,并生成一种 G 代码,其中包含了与某一规格的 3D 打印机(如 FDM 打印机)专门定制的指令。

**语法补充:过去分词做状语的用法**

过去分词作状语可以表示时间(如注释二中的 completed)、地点、原因、条件、让步、伴随等意义。过去分词相当于状语从句,若过去分词作状语,句子的主语与分词所表示的动作构成动宾关系,即是该分词动作的承受者。如:

1) When <u>China Pavilion is seen</u> from the distance, it looks like a Chinese crown.

   = When <u>seen</u> from the distance, China Pabilion looks like a Chinese crown. (时间)

2) Because <u>the trees were decorated</u> with colorful lights, they made an excellent impression on us.

   = Because <u>decorated</u> with colorful lights, they made an excellent impression on us. (原因)

3) If <u>we are given</u> a chance to go on a trip by railcar, called "tour of wisdom", we should treasure it.

   = If <u>given</u> a chance to go on a trip by railcar, called "tour of wisdom", we should treasure it. (条件)

4) We entered Section 3-low carbon exhibition, <u>as we were followed</u> by a group of beautiful girl models.

   = <u>Followed</u> by a group of beautiful girl models, We entered Section 3-low carbon exhibition. (伴随)

5) Though <u>we were exhausted</u>, we spent a meaningful and unforgettable time in it.

   = Though <u>exhausted</u>, we spent a meaningful and unforgettable time in it. (让步)

3. Traditional techniques like injection moulding can be less expensive for manufacturing

polymer products in high quantities, but additive manufacturing can be faster, more flexible and less expensive when producing relatively small quantities of parts.

像注塑成型这样的传统技术,可以花费不高地大批量生产聚合物产品,但增材制造用于生产相对小批量的产品时,速度更快、更灵活而且费用也更少。

4. Selective laser melting (SLM) does not use sintering for the fusion of powder granules but will completely melt the powder using a high-energy laser to create fully dense materials in a layer-wise method that has mechanical properties similar to those of conventional manufactured metals.

选择性激光熔化(SLM)技术不是烧结粉末颗粒进行融合,而是应用高能激光通过一种分层方法,创建出完全致密的材料。它拥有的机械性能与传统制造的金属相当。

## Check your understanding

Ⅰ. Give brief answers to the following questions
    1. What is 3D printing?
    2. Which file type of the object instructs the 3D printing process?

Ⅱ. Match the items listed in the following two columns.

| | |
|---|---|
| additive manufacturing | 分辨率 |
| fixup | 切片 |
| fusing | 颗粒 |
| slicer | 熔合 |
| sintering | 修补 |
| granules | 增材制造 |
| resolution | 制造 |
| fabrication | 烧结 |

# 3D 打印技术

    3D 打印,又称增材制造(AM),指的是堆积形成一个三维产品的各种过程的总称。在 3D 打印过程中,在电脑控制下一层层的材料连续不断地叠加成型,形成了一个实物。这些实物可以是任意的形状或几何体,根据一个 3D 模型或是其他电子数据源被制造出来。一台 3D 打印机就是工业机器人的一种类型。

    未来学家,如 Jeremy Rifkin 等,他们认为 3D 打印是第三次工业革命开端的标志,是开始于19世纪后期主导制造业的生产线装配制造业的延续。借助于互联网的功能，它有可能最终实现如下场景，即将任何一个产品的设计图发送到全世界的任何一个地方，都能使用一台3D打印机将产品复制出来，打印机配有"元素墨水"，有能力将其组合成任意所需形式的任何物体。

3D打印在字面意义上指的是使用喷墨打印机喷头，有序地把材料沉积到粉床上面。近来，此术语的涵义已经拓展和包括更多种类型的技术，诸如挤出和基于烧结的过程。技术标准通常用增材制造描述其更宽泛的意义。

图8.1-1 增材制造

### 1. 概述

用于3D打印的模型可以由计算机辅助设计（CAD）软件包、一台3D扫描仪或者是一个普通的数码相机及摄影测量学软件生成。CAD技术形成的3D打印模型能促使错误减少，可在打印工作之前被更改，实现了打印之前允许对产品的设计进行验证的结果。

为三维计算机图像准备的几何数据的手工建模过程就如同对塑料艺术品完成的雕刻过程。3D扫描就是对一个真实产品的形状和表观进行数字数据的采集过程，最终形成了一个基于实物的数字模型。

在用一个STL文件进行3D打印模型之前，首先必须对数据模型进行多方面的错误检查。这个步骤通常被称为"修补"。一般而言，通过3D扫描获取模型后产生的STL数据经常会存在许多方面的错误，这些错误需要进行修复。这些错误中最典型的实例就是曲面没有相互连接或者在模型中有裂缝。

一旦完成，STL文件需要由一款被称为"切片"的软件进行后处理。这种软件能够把模型转化为一系列的薄片层，并生成一种G代码，其中包含了与某一规格的3D打印机（如FDM打印机）专门定制的指令。这种G代码文件可用于打印，它通过3D打印客户端软件加载G代码，并利用代码指导3D打印机的三维打印加工过程。

打印机的分辨率是由分层厚度和X-Y方向上的分辨率（即每英寸或者每微米的点数）描述。典型的分层厚度是100 μm（250 DPI）左右，虽然有些打印机可以打印薄至16 μm（1,600 DPI）层厚。X-Y方向上的分辨率可以与激光打印机的分辨率相比。微粒（3D点阵）的直径大约在50～100 μm（510～250 DPI）范围。

应用当代的各种方法可以在任何地方制

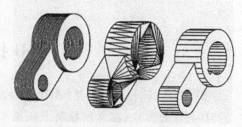

图8.1-2 产品模型
(a)实体模型 (b)一个STL文件 (c)切片分层

作一个模型,花费的时间从数小时到数天。这取决于所使用的方法、模型的大小和复杂程度。增材系统可以缩短这个时间至数个小时,尽管时间的改变很大地取决于所用机器的类型、同时产生模型的规模和数量。

像注塑成型这样的传统技术,可以花费不高地大批量生产聚合物产品,但增材制造用于生产相对小批量的产品时,速度更快、更灵活而且费用也更少。3D打印机为设计师和概念开发团队提供了使用一个桌面大小的打印机就可以生产零部件和概念模型的能力。

打印机实现的分辨率对许多应用是足够的。以标准分辨率打印略微超大版本的所需产品对象,然后进行具有更高分辨率的减材过程,通过去除材料可以获得更高的精度。一些可打印的聚合物使用化学气相过程,能够实现光滑的表面和改善表面精度。

一些增材制造技术在构建零件时能够使用一系列材料。这些技术能够同时打印出多种色彩和颜色组合,从而不需要后期的涂色。

某些打印技术在制造中需要创建内部支撑体,用于实现突出的特征。这些支撑体必须在打印结束后被机械方式移除或被溶解。

## 2. 3D 打印工艺

自 1970 年代末起,人类已经发明了一些 3D 打印工艺过程。最初的打印机体积庞大、价格昂贵、能生产的产品也受到很大程度的限制。

3D 打印技术中的一种方法是在一个颗粒台上有所选择地对材料进行融合。这种技术融合相关的分层,然后在工作区域内向上移动,添加另外一层颗粒,不断重复这个过程,直至工件已经构建成功。这个过程使用不能融合的媒介材料,用来支撑正在成型的零件上的悬臂和薄壁结构,这样可以减少对工件暂时性的辅助支撑装置的需求。

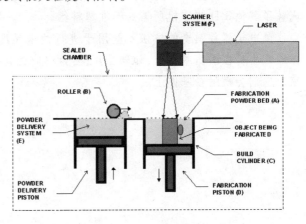

图 8.1-3 选择性激光烧结

一种激光被广泛地用来将媒介烧结成固体。其实例包括对金属和聚合物都适用的选择性激光烧结(SLS)和直接针对金属的激光烧结(DMLS)技术。

选择性激光熔化(SLM)技术不是烧结粉末颗粒进行融合,而是应用高能激光通过一种分层方法,创建出完全致密的材料。它拥有的机械性能与传统制造的金属相似。

另一种打印方法由一台 3D 喷墨打印系统构成。打印机在零件的横截面上,通过喷涂一层粉末(石膏或树脂)和打印上粘合剂,整个过程就像喷墨打印过程一样,每一次创建完成模型的一层。重复此过程,直到每一层都打印出来。这种技术允许打印全彩的原型、悬臂结构、弹性体零件。粘结粉打印的强度可用蜡或热固性聚合物浸渍获得增强。

### 3. 3D 打印机

截至 2012 年 10 月,增材制造系统在市场上的价格从 2,000 美元到 500,000 美元不等,应用的行业包括航空航天、建筑、汽车、国防、医疗替代和许多其他行业。例如,通用电气使用高端模型使涡轮机上的零件成型。很多这样的系统被应用于大规模制造前的快速成型。

随着 3D 打印机费用的下降,3D 打印正越来越吸引人们为此消费,用于自己制作个人的产品。此外,在家里进行 3D 打印,减少了材料的使用和分配的影响,进而会降低对环境的影响

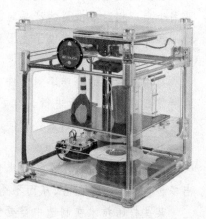

图 8.1-4　3DTouch™
3D 打印机 - 双喷头

已研制出大型的 3D 打印机用于工业、教育和商业的应用。SeeMeCNC 公司在 2014 年曾制造了一台大型 delta-style 3D 打印机。这台打印机能够打印直径至 4 英尺 (1.2 米)、高度达到 10 英尺(3.0 米)的一个物体。它还可以使用塑料颗粒作为原料,代替了其他 3D 打印机所需的标准塑料丝。

微电子器件制造的方法可被用于进行纳米尺度实物的 3D 打印。此类的打印实体通常生长在固体基体上,如硅晶片;因为它们极其微小和脆弱而无法被操纵,打印结束后它们会吸附在基体上。

# Lesson 2  "Internet +" & "Intelligent manufacturing"

1　At present, "Internet +" has become a hot word. The Internet and manufacturing marriage has created an intelligent tide. In "Made in China 2025", the intelligent manufacturing is listed as the main direction and an important path for the practice of the "Internet +" strategy.

## Overview

2　On May 19, 2015, the State Council issued the "Made in China 2025", which is the first 10-year action programme to achieve China's manufacturing power. "Made in China 2025" puts forward that the strategic objectives of manufacturing power can be achieved by the "three-step": first, by 2025 China shall be ranked among manufacturing powers; second, by 2035 China's manufacturing industry as a whole shall reach the moderate level in the world's manufacturing power camp; third, by the 100th anniversary of new China, China's status as a big manufacturing power shall be more consolidated and its comprehensive strength shall be ranked at the forefront among world manufacturing powers.

3　The "Made in China 2025" stressed the need to deepen the application of Internet in the field of manufacturing, to implement industrial cloud and big data innovation pilot, to construct a number of high-quality industrial cloud services and big data platforms, to promote the opening and sharing of software and services, design and manufacturing resources, key technologies and standards.

## "Internet +" opens "Chinese wisdom made"

4　China is currently in the late industrialization and also in the information phase. The "Internet +" mode actually integrates industrialization and informatization organically, uses informatization to transform industrialization, pushes the development of new industries through re-industrialization. "Internet +" mode also helps solve some problems existing in previous manufacturing production where blind production leads to overcapacity easily.

5　The "Internet +" model can also help companies understand the market in real time and organize production and sales based on the information obtained, which helps solve overcapacity.

*6*   "Made in China 2025" also insists on accelerating the integration of the new generation of information technology and manufacturing technology with intelligent manufacturing as the main direction. China's manufacturing industry is facing many constraints, the most obvious one of which is the rising labor costs. The intelligent manufacturing will help to extend the competitive advantage of China's manufacturing industry. The development of intelligent manufacturing can improve productivity, help enterprises to deal with rising labor costs and other issues in the future. Meanwhile it can speed up the standardization of production and maintain a competitive advantage.

**Economic transformation depends on the industrial wisdom made**

*7*   On December 17th, 2015, at the second session of the World Internet Conference held in Wuzhen, The "Internet +" forum launched a discussion on "Intelligent Manufacturing Transformation and Upgrade". Zhejiang Governor Li Qiang the first lecturer, made a keynote speech, and discussed the way of intelligent manufacturing development in the Internet environment with Zhang Ruimin, Chairman of the Board of Directors of Haier Group, Li Shufu, Chairman of Geely Group and other business executives.

*8*   Li Qiang said in his speech, "The new industrial revolution, which is based on a new generation of information technology with 'Internet +' manufacturing as the trend, is a strategic opportunity for 'Made in China' to get rid of 'two-way squeeze'; the deep integration of informatization and industrialization, which makes intelligent manufacturing as the main direction, is the best path for China's industrial economy to realize 'corner overtaking'."

**"Internet + manufacturing": Intelligent manufacturing**

*9*   It is reported that "China 2025" presents nine strategic tasks, five key projects and a number of major policy initiatives, deploys ten strategic areas where key breakthroughs should be achieved, such as the new generation of information technology industry, high-end CNC machine tools and robots, aerospace equipment, marine engineering equipment and high-tech ships, advanced rail transportation equipment, energy-saving and new energy vehicles, electrical power equipment, agricultural equipment, new materials, bio-medicine and high-performance medical equipment.

*10*   It is worth noting that it is the main line of "China 2025" to accelerate the integration of the new generation of information technology and manufacturing. Some analysts believe that in recent years, the internet of things, cloud computing and big data have become important parts of the new generation of information technology. With the advent of a new round of industrial revolution, they will also play an increasingly important role in the manufacturing sector.

11   "Made in China 2025" focuses on the development of emerging technologies which is closely related to the previously determined seven strategic emerging industries. It is the main line of promoting "Made in China 2025" to treat the "Internet + manufacturing" as the core, vigorously develop intelligent manufacturing and push the deep integration of informatization and industrialization.

## ❖ New Words and Phrases

| | |
|---|---|
| Made in China 2025 | 中国制造2025 |
| strategy[ˈstrætədʒi] n. | 战略,策略 |
| the State Council | 国务院 |
| tide[taid] n. | 潮流,趋势 |
| implement[ˈImplImənt] v. | 实施,执行 |
| put forward | 提出 |
| rank[ræŋk] v. | 排列,列为 |
| power[ˈpauə] n. | 强国,大国 |
| consolidate[kənˈsɔlideit] v. | 巩固,加强 |
| forefront[ˈfɔːfrʌnt] n. | 前列,前沿,最前方 |
| innovation pilot | 创新试点 |
| platform[ˈplætfɔːm] n. | 平台,台,站台 |
| Chinese wisdom made | 中国智造 |
| organically[ɔːˈgænikli] adv. | 有机地 |
| overcapacity[ˌəuvəkəˈpæsəti] n. | 生产能力过剩 |
| generation[ˌdʒenəˈreiʃən] n. | 代,时代,产生 |
| integration[ˌintiˈgreiʃən] n. | 融合,结合,一体化 |
| constraint[kənˈstreint] n. | 限制,约束 |
| economic transformation | 经济转型 |
| forum[ˈfɔːrəm] n. | 论坛 |
| keynote[ˈkiːnəut] n. | 主旨,重点,基调 |
| squeeze[skwiːz] n. | 挤压,压榨 |
| corner overtaking | 弯道超车 |
| initiative[iˈniʃətiv] n. | 倡议,主动性(权),主动精神 |
| deploy[diˈplɔi] v. | 部署,调度,调配 |
| bio-medicine[ˌbaiəuˈmedIsn] n. | 生物医药 |

sector [ˈsektə(r)] n.     部门,领域
emerging technology     新兴技术
related [rɪˈleɪtɪd] adj.     相关的

## ❖ Notes

1. "Internet +" has become a hot word. The Internet and Manufacturing marriage has created an intelligent tide.

    "互联网+"成为一大热词。互联网与制造业联姻更是开启了智能化大潮。

2. On May 19, 2015, the State Council issued the "Made in China 2025", which is the first 10-year action programme to achieve China's manufacturing power.

    2015年5月19日,国务院印发《中国制造2025》,这是中国实施制造强国第一个十年的行动纲领。

3. "Internet +" mode also helps solve some problems existing in previous manufacturing production where blind production leads to overcapacity easily.

    "互联网+"的模式也利于解决以前制造业生产存在的一些问题,以前由于生产比较盲目,易导致产能过剩。

4. "Made in China 2025" also insists on accelerating the integration of the new generation of information technology and manufacturing technology with intelligent manufacturing as the main direction.

    《中国制造2025》还明确,加快推动新一代信息技术与制造技术融合发展,把智能制造作为两化深度融合的主攻方向。

## 语法补充:insist 的用法

1. insist on/upon (sb/one's) (doing) sth:坚持(做)某事如:

You should insist on/upon your dream. 你应该坚持你的梦想。

You should insist upon/on doing exercise every day. 你应该每天坚持锻炼。

You should insist on/upon her apologizing to you. 你应该坚持让她向你道歉。

2. insist + that 引导的宾语从句

1) 如果 insist 翻译为"坚持说/坚持认为",即 that 从句表示已经发生的动作时,则宾语从句使用陈述语气,即从句该用什么时态就用什么时态,当然还要和主句时态保持呼应。如:

She insisted (that) she was right. 她坚持认为自己是对的。

2) 如果 insist 翻译为"坚持要求/坚持主张",即 that 从句的动作当时尚未发生的话,则宾语从句必须使用虚拟语气,即从句谓语动词由"should + 动词原形"构成,且 should 可以省略。如:

He insisted that she (should) say sorry to him first. 他坚持要求她先向他道歉。

5. It is worth noting that it is the main line of "China 2025" to accelerate the integration of the new generation of information technology and manufacturing.

值得注意的是,加快新一代信息技术与制造业的融合成为《中国制造2025》的主线。

## Check your understanding

Ⅰ. Give brief answers to the following questions:
1. What is "Made in China 2025"?
2. What is put forward in "Made in China 2025" strategy?
3. What is the development trend of manufacturing?

Ⅱ. Match the items listed in the following two columns:

| | |
|---|---|
| integration | 主旨 |
| keynote | 前列 |
| forefront | 实施 |
| implement | 融合 |
| Internet plus | 弯道超车 |
| innovation pilot | 新兴产业 |
| corner overtaking | 中国智造 |
| emerging industry | 互联网+ |
| Chinese wisdom made | 创新试点 |

# "互联网+"与智能制造

目前"互联网+"成为一大热词。互联网与制造业联姻更是开启了智能化大潮。在《中国制造2025》中,智能制造被列为主攻方向,也是实践"互联网+"战略的重要路径。

## 概述

2015年5月19日,国务院印发《中国制造2025》,这是中国实施制造强国第一个十年的行动纲领。《中国制造2025》提出,通过"三步走"实现制造强国的战略目标:第一步,到2025年迈入制造强国行列;第二步,到2035年我国制造业整体达到世界制造强国阵营中等水平;第三步,到新中国成立一百年时,我国制造业大国地位更加巩固,综合实力进入世界制造强国前列。

《中国制造2025》强调,要深化互联网在制造领域的应用,实施工业云及工业大数据创新应用试点,建设一批高质量的工业云服务和工业大数据平台,推动软件与服务、设计与制造资源、关键技术与标准的开放共享。

### "互联网+"开启"中国智造"

目前我国正处在工业化后期,而且也正处在信息化阶段,"互联网+"模式实际上就是把工业化和信息化有机地融合在一起,用信息化改造工业化,通过再工业化推动新兴产业的发展。"互联网+"的模式也利于解决以前制造业生产存在的一些问题,以前由于生产比较盲目,易导致产能过剩。

"互联网+"的模式还可以帮助企业实时了解市场行情,根据所获信息,组织安排生产、销售,提高生产效率,也有利于解决产能过剩问题。

《中国制造2025》还明确,加快推动新一代信息技术与制造技术融合发展,把智能制造作为两化深度融合的主攻方向。我国制造业发展面临不少制约,其中最明显的就是劳动力成本上升。而智能制造有利于延长我国制造业的竞争优势,发展智能制造可以提高生产效率,有利于企业应对未来劳动力成本上升等问题,同时也可以加快标准化生产,保持竞争优势。

### 经济转型依赖产业智造

2015年12月17日,在乌镇举行的第二届世界互联网大会上,"互联网+"论坛发起"智能制造转型与升级"议题讨论。浙江省省长李强首先作主旨演讲,并与海尔集团董事局主席张瑞敏、吉利集团董事长李书福等企业负责人共同探讨互联网环境下的智能制造发展之路。

李强在演讲中指出,"以新一代信息技术为基础,以'互联网+'制造为趋势的新工业革命,是中国制造摆脱'双向挤压'的战略机遇;以智能制造为主攻方向的信息化和工业化的深度融合,是中国工业经济'弯道超车'的最佳路径。"

### "互联网+制造业":智能制造

《中国制造2025》提出了九大战略任务、五项重点工程和若干重大政策举措,部署了重点突破的十大战略领域,比如新一代信息技术产业、高档数控机床和机器人、航空航天装备、海洋工程装备及高技术船舶、先进轨道交通装备、节能与新能源汽车、电力装备、农机装备、新材料、生物医药及高性能医疗器械等。

值得注意的是,加快新一代信息技术与制造业的融合成为《中国制造2025》的主线。有分析认为,近几年来,物联网、云计算、大数据成为新一代信息技术的重要内容,随着新一轮工业革命来临,它们在制造业中也将发挥愈加重要的作用。

《中国制造2025》重点发展的新兴技术,与此前确定的七大战略性新兴产业相关度较高。以"互联网＋制造业"为核心,大力发展智能制造,推进信息化与工业化深度融合,是推进《中国制造2025》的主线。

**构词法之截短法**
一、定义：截短法指将单词缩写,词义和词性保持不变。
二、具体构成方式
（一）截头
telephone（[名词] 电话）→ phone（[名词] 电话）
airplane（[名词] 飞机）→ plane（[名词] 飞机）
（二）去尾
mathematics（[名词] 数学）→maths（[名词] 数学）
examination（[名词] 考试）→exam（[名词] 考试）
kilogram（[名词] 千克）→kilo（[名词] 千克）
laboratory（[名词] 实验室）→lab（[名词] 实验室）
taxicab（[名词] 出租车）→taxi（[名词] 出租车）
3. 截头去尾
influenza（[名词] 流行性感冒）→flu（[名词] 流行性感冒）
refrigerator（[名词] 电冰箱）→fridge（[名词] 电冰箱）

# Answer to Exercises

## The History of Metallurgy

### Check your understanding

Ⅰ. Give brief answers to the following questions.
1. Who is called the "father of metallurgy"?
   Agricola has been described as the "father of metallurgy".
   谁被称为"冶金之父"? 答:阿格里科拉被称为"冶金之父"。

Ⅱ. Match the items listed in the following two columns.
trip hammer 杵锤   compound 混合物   blast furnace 鼓风炉,高炉   cast iron 生铁   meteoric iron 陨铁   metalworking 金属加工   metallurgy 冶金;冶金学

## Metallurgy

### Check your understanding

Ⅰ. Give brief answers to the following questions.
1. What is metallurgy?
   Metallurgy is a domain of materials science that studies the physical and chemical behavior of metallic elements, their intermetallic compounds, and their mixtures, which are called alloys.
   什么是冶金学? 答:冶金学是研究金属元素、金属间化合物及其混合物(即合金)的物理和化学特性的科学,也是金属工艺学,生产中的科学。
2. How are metals shaped?
   Metals are shaped by processes such as casting, forging, flow forming, rolling, extrusion, sintering, metalworking, machining and fabrication.
   金属成形工艺有哪些? 答:金属成形工艺有铸造、锻造、流动旋压、轧制、挤压、烧结,金属加工,机械加工和制作。
3. What is welding?
   Welding is a technique for joining metal components cohesively by melting the base material, making the parts into a single piece.
   什么是焊接? 答:焊接是通过加热使分离的物体连接成整体的技术。

Ⅱ. Match the items listed in the following two columns.
toughness 韧性   extractive metallurgy 冶炼   plating 电镀   rolling 轧制   drill 钻床   welding 焊接   guillotine 轧刀   microstructure 微观结构   lathe 车床

# Metals

**Check your understanding**

Ⅰ. Give brief answers to the following questions.

1. What are the five most used metals?

    Iron, aluminum, copper, zinc and magnesium.

    有哪五种最常用的金属?答:铁、铝、铜、锌、镁。

2. Which metal is used for electrical cables?

    Copper is used for electrical cables.

    哪种金属适用于电缆?答:铜适用于电缆。

3. What features does aluminum have?

    Aluminum is a good conductor of heat and is malleable. It is used to make saucepans and tin foil, and also aeroplane bodies as it is very light.

    铝有什么样的特点?答:铝具有良好的导热性和可塑性,常用来制造锅和锡箔;因为它很轻,因此也用来制造飞机的部件。

Ⅱ. Match the items listed in the following two columns.

zinc　锌　heat conductor　导热体　pressure vessel　压力容器　inconel　铬镍铁合金　alloy　合金　stainless steel　不锈钢　galvanized steel　镀锌钢　magnetic　有磁性的

# Materials in industry

**Check your understanding**

Ⅰ. Give brief answers to the following questions.

1. What are the industrial applications of materials science?

    Industrial applications of materials science include materials design, cost-benefit tradeoffs in industrial production of materials, processing techniques and analytical techniques, etc.

    材料科学的工业应用包括哪些?答:材料科学的工业应用包括材料设计,工业生产中成本效益权衡,加工工艺和分析技术等等。

2. For the steels, what directly affects the hardness and tensile strength of the steel?

    For the steels, the hardness and tensile strength of the steel is directly related to the amount of carbon present.

    对于钢而言,什么直接影响钢的硬度和拉伸强度?答:对于钢而言,含碳量直接影响钢的硬度和拉伸强度。

3. What are the composite materials?

The composite materials are structured materials composed of two or more macroscopic phases.

什么是复合材料？答：复合材料是由两种或两种以上的宏观相组成的结构材料。

Ⅱ. Match the items listed in the following two columns.

ion　离子　polyethylene　聚乙烯　electrolytic extraction　电解提取　nickel　镍
nylon　尼龙　titanium　钛　dispersant　分散剂　polyester　聚酯纤维，涤纶

# Casting

## Check your understanding

Ⅰ. Give brief answers to the following questions.

1. What is casting?

   Casting is a manufacturing process by which a liquid material is (usually) poured into a mold, which contains a hollow cavity of the desired shape, and then allowed to solidify.

   什么是铸造？答：铸造是将液体浇进铸型里，铸型是空心的，然后凝固成形得到预定形状的一种制造工艺过程。

2. Which two distinct subgroups is the casting process subdivided into?

   Expendable and non-expendable mold casting.

   铸造工艺可以分为哪两种？答：消失模铸造和非消失性模铸造。

3. Compared to permanent mold casting, what features does sand casting have?

   Not only does this method allow manufacturers to create products at a low cost, but there are other benefits to sand casting, such as very small size operations. Sand casting also allows most metals to be cast depending on the type of sand used for the molds.

   砂型铸造与金属型铸造相比具有哪些特点？答：适用于小批量、低成本生产；并且工作范围小，可以通过不同种类砂子的模具铸造大多数金属。

4. What fields is shell molding applied in?

   Shell molding is used for small parts that require high precision, such as gear housings, cylinder heads and connecting rods. It is also used to make high-precision moulding cores.

   壳模应用在哪些方面？答：壳模适用于精度要求高的小零件，例如齿轮外壳、气缸盖和连杆。它同样也可用来制造高精度的铸模砂心。

Ⅱ. Match the items listed in the following two columns.

casting　铸造　mold　铸　gypsum　石膏　alloy　合金　ferrous material　黑色金属
　plaster casting　石膏模铸造　investment casting　熔模铸造　component　部件
cavity　腔　gear　齿轮

# Forging

**Check your understanding**

Ⅰ. Give brief answers to the following questions.

1. What may be forging classified into?
   Cold forging, hot forging and warm forging.
   锻造一般分为哪几种？答：有冷锻，热锻和温锻。

2. What is the advantage of hot forging?
   The main advantage of hot forging is that as the metal is deformed the strain-hardening effects are negated by the recrystallization process.
   热锻有什么优点？答：热锻的主要优点是在再结晶过程中不会发生应变硬化效应，便于金属成形。

3. What do common forging processes include?
   Roll forging, swaging, cogging, open-die forging, impression-die forging, press forging, automatic hot forging and upsetting.
   一般锻造工艺包括什么？答：辊锻备坯、模锻成形、锻造开坯、自由锻、开式模锻、锻压、自动化热锻和镦粗。

4. What is the most common type of forging equipment? The most common type of forging equipment is the hammer and anvil.
   最常见的锻压设备是什么？答：最常见的锻压设备是铁锤和铁砧。

Ⅱ. Match the items listed in the following two columns.

forging 锻造　recrystallization 再结晶　forging press 锻压机　hydraulic press 液压机　metalworking 金属加工　cogging 开坯　roll forging 辊锻　cam 凸轮　connecting rod 连杆　steam hammer 蒸汽锤

# Welding

**Check your understanding**

Ⅰ. Give brief answers to the following questions.

1. What is welding?
   Welding is a fabrication or sculptural process that joins materials, usually metals or thermoplastics, by causing coalescence.
   什么是焊接？答：焊接是一种制作和雕刻工艺，它可以使材料，通常是金属或热塑性塑料之间合并连接起来。

2. What is the most common type of arc welding？
   One of the most common types of arc welding is shielded metal arc welding (SMAW),

which is also known as manual metal arc welding (MMA) or stick welding.

最常见的电弧焊是什么？答：最常见的电弧焊是焊条电弧焊（SMAW），也称为手工金属电弧焊（MMA）或焊条焊接。

Ⅱ. Match the items listed in the following two columns.

arc welding　电弧焊　resistance welding　电阻焊　forge welding　锻焊　submerged arc welding　埋弧焊　stud arc welding　螺栓电弧焊　oxyfuel welding　气焊　flux cored arc welding　药芯焊丝电弧焊　laser beam welding　激光焊

## Lathe (metal)

**Check your understanding**

Ⅰ. Give brief answers to the following questions.

1. At the least, what do mental lathes consist of?

   Mental lathes consist of, at the least, a headstock, bed, carriage and tailstock.

   金属车床的组成至少包括哪些？答：金属车床的组成至少包括主轴箱，床身，拖板箱和尾架。

2. What is the function of the bed?

   The bed connects to the headstock and permits the carriage and tailstock to be aligned parallel with the axis of the spindle.

   床身的作用是什么？答：床身连接主轴箱并使拖板箱和尾座沿着床身上的导轨作平形于主轴方向的直线运动。

Ⅱ. Match the items listed in the following two columns.

toolroom lathe　工具车床　fine feed　微小进给　boring　钻孔　quadrant plate　挂轮板　tumbler gear　摆动换向齿轮　tool post　刀架，刀座　chuck　卡盘　driveshaft　驱动杆　pulley　皮带轮　feedscrew　进给螺杆

## Milling Machine

**Check your understanding**

Ⅰ. Give brief answers to the following questions.

1. What is the basic form of a milling machine?

   The basic form of a milling machine is that of a rotating cutter which rotates about the spindle axis, and a table to which the workpiece is affixed.

   铣床的基本组成有哪些？答：铣床的基本组成有旋转刀具（绕着轴旋转，就像钻头）和用来固定住工件的工作台。

2. What are the subcategories of vertical mills?

   There are two subcategories of vertical mills: the bedmill and the turret mill.

# Answer to Exercises

立式铣床可分为哪两种主要的类型？答：立式铣床有两种主要的类型：床式和转塔式。

Ⅱ. Match the items listed in the following two columns.

the turret mill　转塔式铣床　knee mill　膝型铣床　crane　起重机　floor mill　落地铣床　jig borer　工模镗孔机　bed mill　床式铣床　ram type mill　维修型铣床　column mill　圆柱形铣床　box mill　盒型铣床

## Types of milling Cutter

### Check your understanding

Ⅰ. Give brief answers to the following questions.

1. What does the end mill include?

   The end mill generally refers to flat bottomed cutters, but also include rounded cutters and radiused cutters.

   立铣刀包括哪些？答：立铣刀这个称呼通常指平底铣刀，但也包括圆周切削铣刀和径向铣刀。

2. What are the functions of the woodruff cutters?

   Woodruff cutters make the seat for woodruff keys. These keys retain pulleys on shafts and are shaped as shown in the image.

   半圆键铣刀的用途是什么？答：半圆键铣刀用来加工半圆键，这些键用于滑轮和传动轴之间的连接。

Ⅱ. Match the items listed in the following two columns.

rounded cutters　圆周切削铣刀　screw machines　螺纹加工机床　radiused cutters　径向切削铣刀　keyway slots　键槽　slot drills　键槽铣刀　roughing end mills　粗加工端铣刀　carbide tipped face mill　硬质合金刀片面铣刀　cylindrical boss　圆柱形工件　woodruff key cutters　半圆键铣刀

## Engine Operating Principles

### Check your understanding

Ⅰ. Give brief answers to the following questions.

1. What are most automobile engines?

   Most automobile engines are internal combustion, reciprocating 4-stroke gasoline engines.

   大多数汽车用什么样的发动机？答：大多数汽车用发动机都是内部燃烧、往复运动四冲程汽油发动机。

2. What is called the engine bore?

The diameter of the cylinder is called the engine bore.

什么叫发动机内径? 答:气缸的直径称为发动机内径。

Ⅱ. Match the items listed in the following two columns.

throttle　节气门　oil pan　油底壳　combustion　燃烧　bearing　轴承　displacement　排气量　compression ratio　压缩比　piston pin　活塞销　exhaust stroke　排气行程

## Engine Construction

**Check your understanding**

Ⅰ. Give brief answers to the following questions.

1. What is the largest and the most complicated single piece of metal in the automobile?

   Usually, The cylinder block is the largest and the most complicated single piece of metal in the automobile.

   什么是汽车中最大最复杂的独立金属件? 答:通常,气缸是汽车中最大最复杂的独立金属件。

2. What is the function of piston?

   The piston converts the potential energy of the fuel into the kinetic energy that turns the crankshaft.

   活塞的作用是什么? 答:活塞将燃油的潜在能量转化为动能来驱动曲轴。

Ⅱ. Match the items listed in the following two columns.

bearing cap　轴承盖　rotary motion　旋转运动　automobile　汽车、车辆　finished block　成品机体　kinetic energy　动能　water jackets　水套　gray iron　灰铸铁　crankshaft　机轴

## Forming of Sheet Metals

**Check your understanding**

Ⅰ. Give brief answers to the following questions.

1. What is metalworking?

   Metalworking is the process of working with metals to create individual parts, assemblies, or large scale structures.

   什么是金属加工? 答:金属加工是对加工金属材料,从而生产出独立的零件、装配体或者是规模很大的结构件的过程。

2. What is metalworking generally divided into?

   Metalworking generally is divided into the following categories, forming, cutting, and, joining.

   金属加工分成哪几类? 答:金属加工从总体上分成以下几类:金属成形,切割和连

# Answer to Exercises

结。

Ⅱ. Match the items listed in the following two columns.

stamping 冲压件 sheet metal 金属片 metal stamping dies 金属冲压模具 plating 电镀 exotic metals 特殊金属 accuracy 准确性 stability 稳定性 jewellery 珠宝

## Metal Stamping Process and Die Design

### Check your understanding

Ⅰ. Give brief answers to the following questions.

1. What is bending?

   The bending operation is the act of bending blanks at a predetermined angle.

   什么是弯曲？答：弯曲工序是一种将坯料弯曲成预定角度的冲压工序。

2. What does a die set consist of?

   A die set consists of a lower shoe and an upper shoe.

   模架包括什么？答：模架包括下模座和上模座。

Ⅱ. Match the items listed in the following two columns.

stationary stripper 固定式卸料板 die set 模架 piercing 冲孔加工 side gauges 侧压板 lower shoe 下模座 die buttons 冲模凹模 guide pins 导柱 back-up plates 支承垫板 hardened-tool steel 淬硬工具钢

## Plastics and Injection Molding

### Check your understanding

Ⅰ. Give brief answers to the following questions.

1. What are plastics composed of?

   Plastics are composed of polymer molecules and various additives.

   塑料的组成成分是什么？答：塑料的组成成分是高聚物分子和各种添加剂。

2. What are the features of plastics?

   Plastics are characterized by the following properties: low density, low strength and elastic modulus, low thermal and electrical conductivity, high chemical resistance, and high coefficient of thermal expansion.

   塑料具有什么特点？答：塑料具有如下特点：密度低、强度和弹性模量小，导电和导热性能低、化学稳定性和热膨胀系数高。

Ⅱ. Match the items listed in the following two columns.

elastomer 弹性体 the clamping unit 锁模单元 elastic modulus 弹性模量 injection molding 注射成型 polymer 聚合体 thermal conductivity 热传导率

thermosets 热固性塑料　mold 模具　plasticating 塑化

## Design of Injection Mold

**Check your understanding**

Ⅰ. Give brief answers to the following questions.

1. What is the function of a mold?
   The function of a mold is twofold: imparting the desired shape to the plasticized melt and solidifying the injected molded product (cooling for thermoplastics and heating for thermoset plastics).
   注射模具的作用是什么？答：注射模具的作用有两个方面：使充分塑化的塑料熔体形成特定的形状，对注射模塑的产品进行固化（热塑性塑料采用冷却方式，热固性塑料采用加热方式）。

2. What two basic parts does the mold have?
   The mold has two basic parts to contain the cavities and cores. They are the stationary mold half on the side where the plastic is injected, and a moving half on the closing or ejector side of the machine.
   注射模具分成哪两大部分？答：注射模具分成两大部分容纳凹模和型芯。一半是定模，在注射塑料的一侧；另一半是动模，在注射设备负责闭合和开启模具的一侧。

Ⅱ. Match the items listed in the following two columns.
core 型芯　cavity 型腔,模腔,凹模　mold base 模架,模座　sprue bushing 主流道衬套　nozzle 喷嘴　parting line 分型线　stripper plate 脱料板　sleeve 套筒　gate 浇口

## Food Robotics

**Check your understanding**

Ⅰ. Give brief answers to the following questions.

1. What are the applications of robotics in the food and beverage industry?
   For "traditional" applications such as picking, packing and palletizing; as well as for cutting-edge applications such as meat cutting and beverage dispensing.
   机器人在食品和饮料行业的应用有哪些？答：有分拣、包装和码垛这些传统的应用，以及类切割和饮料配送等较前沿的应用。

2. What is palletizing? Palletizing has robots putting the cases or cartons that contain packaged foods onto a shipping pallet.
   什么是码垛？答：码垛就是让机器人把装有包装好的食品箱子或纸盒子放到货盘上。

Ⅱ. Match the items listed in the following two columns.
dispense 分发,分配   retailer 零售商   automation 自动化   wrap 包,裹   raw food 生食食物   gripper 夹子   dish-making 装盘   roboBar 机器人酒保

# Robotic Inspection

## Check your understanding

Ⅰ. Give brief answers to the following questions.

1. Compared with traditional inspection solutions, what advantages do robotic inspection systems have?
   Robotic inspection systems offer cost savings over traditional inspection solutions.
   与传统的检测方案相比,机器人检测系统有什么优势? 答:机器人检测系统比传统的检测方案节省成本。

Ⅱ. Match the items listed in the following two columns.
work cell 单元式生产   nut 螺母   flaw detection 裂缝检查   material handling 物料输送   oil filter 滤油器   error-proofing 纠错   bolt 螺栓   vision system 视觉系统

# Computer-Aided Design (CAD)

## Check your understanding

Ⅰ. Give brief answers to the following questions.

1. Currently, what system are most CAD systems using? Currently, most CAD systems are using interactive graphics system that is very convenient for beginners.
   当前,大多数计算机辅助设计系统采用的是什么系统方式? 答:当前,大多数计算机辅助设计系统采用的是交互图形系统方式。

2. What is one of the most difficult problems in CAD drawings? One of the most difficult problems in CAD drawings is the elimination of hidden lines.
   在 CAD 绘图中,最主要的难题之一是什么? 答:在 CAD 绘图中,最主要的难题之一就是隐藏线的消除。

Ⅱ. Match the items listed in the following two columns.
diagram 图解,图表   tube 管,软管   Computer-Aided Design 计算机辅助设计   traditional graphics system 非交互图形系统   finite element method 有限元分析方法   interactive graphics system 交互图形系统   artificial 人造的   distribution 分布   solid geometry modeling 实体建模

# Computer aided manufacturing (CAM)

**Check your understanding**

Ⅰ. Give brief answers to the following questions.
1. What is CAM?
   Computer aided manufacturing (CAM) is the use of computer-based software tools that assist engineers and machinists in manufacturing or prototyping product components. CAM is a programming tool that makes it possible to manufacture physical models using computer aided design (CAD) programs. CAM creates real life versions of components designed within a software Package.
   什么是计算机辅助制造？答:计算机辅助制造(CAM)是以计算机软件作为工具,来协助工程师和机械工制造产品部件或其原型。计算机辅助制造是一种编程工具,通过使用计算机辅助设计程序来制造物理原型。通过使用软件包,计算机辅助制造创造了设计零件的立体模型。

Ⅱ. Match the items listed in the following two columns.
machinist　机械师　numerical control　数字控制　visualization tools　可视化工具　precision　精度　CAM packages　计算机辅助制造包　integration　集成　flexibility　柔韧性　computer aided manufacturing　计算机辅助制造

# Computer Numerical Control

**Check your understanding**

Ⅰ. Give brief answers to the following questions.
1. What are the most advanced CNC milling-machines?
   The 5-axis machines are the most advanced CNC milling-machines.
   什么是最先进的数控加工中心？答:5轴联动就是最先进的数控加工中心。
2. What tooling will CNC Milling machines use?
   CNC Milling machines will nearly always use SK (or ISO), CAT, BT or HSK tooling.
   数控机床常使用哪几种刀具？答:数控机床常使用 SK (or ISO), CAT, BT or HSK 几种刀具。

Ⅱ. Match the items listed in the following two columns.
drawbar　挂钩,拉杆　engrave　雕刻　workpiece　工件　spindle　轴　designate　指定,标示　toolholder　刀柄　collet　筒夹,夹头　shank　胫　interchangeability　可交换性

# 3D Printing Technology

Ⅰ. Give brief answers to the following questions

Answer to Exercises

1. What is 3D printing?

   3D printing, also known as additive manufacturing (AM), refers to various processes used to synthesize a three-dimensional object.

   3D打印是什么？3D打印，又称增材制造(AM)，指的是堆积形成一个三维产品的各种过程的总称。

2. Which file type of the object instructs the 3D printing process?

   The G-code file can then be printed with 3D printing client software which loads the G-code, and uses it to instruct the 3D printer during the 3D printing process.

   物体的什么类型的文件能够指导3D打印机的加工过程？G代码文件可用于打印，它通过3D打印客户端软件加载G代码，并利用代码指导3D打印机的三维打印加工过程。

Ⅱ. Match the items listed in the following two columns.

additive manufacturing　增材制造　fixup　修补　fusing　熔合　slicer　切片
sintering　烧结　granules　颗粒　resolution　分辨率　fabrication　制造

# "Internet +" & "Chinese wisdom made"

## Check your understanding

Ⅰ. Give brief answers to the following questions:

1. What is "Made in China 2025"?

   It is the first 10-year action programme to achieve China's manufacturing power.

   什么是《中国制造2025》？《中国制造2025》是中国实施制造强国第一个十年的行动纲领。

2. What is put forward in "Made in China 2025" strategy?

   "Made in China 2025" puts forward that the strategic objectives of manufacturing power can be achieved by the "three-step".

   《中国制造2025》提出了什么策略？《中国制造2025》提出，通过"三步走"实现制造强国的战略目标。

3. What is the development trend of manufacturing?

   The Internet +' manufacturing is the trend.

   制造的发展趋势是什么？制造的发展趋势"互联网+"制造。

Ⅱ. Match the items listed in the following two columns:

Integration　融合　keynote　主旨　forefront　前列　implement　实施　Internet plus　互联网+　corner overtaking　弯道超车　innovation pilot　创新试点　Chinese wisdom made　中国智造　emerging industry　新兴产业

# Vocabulary

## A

accuracy [ˈækjurəsi] n.     精确性
acrylics [əˈkriliks] n.     丙烯酸树脂
acrylonitrile [ˌækrələuˈnaitril] n.     丙烯腈(氰乙烯)
adhere [ədˈhɪə(r)] v.     粘附
adjusted [əˈdʒʌstid] a.     调整过的
advent [ˈædvənt] n.     出现,到来
aligned [əˈlaind] a.     对齐的,均衡的
aligned [əˈlaind] a.     排列的
alloy [ˈælɔi] n.     合金
aluminium [ˌæljuˈminiəm] n.     铝
ambiguously [æmˈbigjuəsli] ad.     含糊不清地
amenable [əˈmiːnəbəl] adj.     易控制的
analogous [əˈnæləgəs] adj.     类似的
annealing [ˈaːniːlin] n.     退火
anvil [ˈænvil] n.     铁砧
appendant [əˈpendənt] n.     下垂物
arbor [ˈaːbə] n.     藤架,凉亭
artificial [ˌaːtiˈfiʃəl] adj.     人造的
assist [əˈsist] v.     辅助
asymmetric [ˌeisiˈmetrik] adj.     不对称的
atop [əˈtɔp] ad.     在顶上
attaching [əˈtætʃiŋ] a.     附属的
automation [ˌɔːtəˈmeiʃən] n.     自动化
automobile [ˈɔːtəməbiːl] n.     汽车、车辆
axis [ˈæksis] n.     轴

## B

bartender [ˈbaːˌtendə] n.     酒吧间男招待
batch [bætʃ] n.     一批,一组

| | |
|---|---|
| beam [biːm] n. | 光线 |
| bearing ['bɛəriŋ] n. | 轴承 |
| binder ['baində] n. | 黏合物 |
| biomaterial [baiəumə'tiəriəl] n. | 生物材料 |
| bio-medicine [ˌbaɪəʊ'medɪsn] n. | 生物医药 |
| blacksmith ['blækˌsmiθ] n. | 铁匠,锻工 |
| blanking ['blæŋkiŋ] n. | 空白,下料,落料 |
| blast [blɑːst] n. | 爆破,冲击波 |
| bolt [bəult] n. | 螺栓 |
| bond [bɔnd] n. | 键　vt. 使黏合,使结合 |
| borer ['bɔːrə] n. | 钻孔器 |
| boring ['bɔːriŋ] n. | 钻孔 |
| bottleneck ['bɔtlˌnek] n. | 瓶颈 |
| brazing ['breiziŋ] n. | 硬钎焊 |
| brittleness ['britlnis] n. | 脆性 |
| bronze [brɔnz] n. | 青铜 |
| burlap ['bəːlæp] n. | 粗麻布 |

## C

| | |
|---|---|
| calibrate ['kælibreit] v. | 校准(刻度,使……标准化,测定) |
| cam [kæm] n. | 凸轮 |
| cappuccino [kæpʊ'tʃiːnəʊ] n. | 热牛奶咖啡 |
| carbide ['kɑːbaid] n. | 碳化物 |
| carburize ['kɑːbjuraiz] vt. | 使渗碳 |
| carcass ['kɑːkəs] n. | (屠宰后)畜体 |
| carriage ['kæridʒ] n. | 溜板,拖板 |
| carve [kɑːv] vt. & vi. | 雕刻 |
| casino [kə'siːnəu] n. | 赌博娱乐场 |
| cast [kɑːst] n. | 铸造 |
| casting ['kɑːstiŋ] n. | 铸造 |
| cathode ['kæθəud] n. | 阴极 |
| cavity ['kæviti] n. | 型腔,模腔,凹模 |
| challenging ['tʃælindʒiŋ] a. | 复杂的 |
| chore [tʃɔː] n. | 家务杂事 |
| chromium ['krəumjəm] n. | 铬 |

| | |
|---|---|
| chuck [tʃʌk] n. | 卡盘 |
| clamp [klæmp] n. | 夹具、锁紧 vt. 夹紧,固定 |
| clay [klei] n. | 泥土 |
| coalescence [ˌkəuə'lesns] n. | 合并,结合,联合 |
| coefficient [ˌkəui'fiʃənt] n. | 系数 |
| cogging ['kɔgiŋ] n. | 开坯 |
| coining ['kɔiniŋ] n. | 精压(立体挤压,压花,压印加工) |
| collet ['kɔlit] n. | 筒夹,夹头 |
| combustion [kəm'bʌstʃən] n. | 燃烧,焚烧 |
| complete [kəm'pli:t] a. | 全部的 |
| component [kəm'pəunənt] n. | 部件,原件 |
| compound ['kɔmpaund] n. | 混合物,化合物 |
| compress [kəm'pres] vt. | 压缩 |
| compression [kəm'preʃ(ə)n] n. | 压缩 |
| concentrate ['kɔnsentreit] v. | 浓缩,富集 |
| conductivity [ˌkɔndʌk'tiviti] n. | 传导性,传导率 |
| conductor [kən'dʌktə] n. | 导体 |
| conical ['kɔnikəl] adj. | 圆锥体的 |
| conjunction [kən'dʒʌŋkʃən] n. | 连接,连合 |
| consolidate [kən'sɔlideit] v. | 巩固,加强 |
| constraint [kən'streint] n. | 限制,约束 |
| contamination [kənˌtæmi'neiʃən] n. | 污染物 |
| contour ['kɔntuə] n. | 轮廓 |
| conversion [kən'və:ʃən] n. | 转变,改变信仰,换位 |
| copper ['kɔpə] n. | 铜 |
| core [kɔ:] n. | 型芯 |
| cored [kɔ:] adj. | 带心的 |
| corrosion [kə'rəuʒən] n. | 腐蚀,受腐蚀的部位 |
| counterblow ['kauntəˌbləu] n. | 对击 |
| craft [krɑ:ft] n. | 工艺 |
| crane [krein] n. | 起重机 |
| crank [kræŋk] n. | 曲柄 |
| crankshaft ['kræŋkʃɑ:ft] n. | 机轴 |
| creep [kri:p] n. | 蠕变 |
| criteria [krai'tiəriə] n. | 标准 |

| | |
|---|---|
| critical ['kritikəl] adj. | 关键性的 |
| cross-section ['krɒs'sekʃən] n. | 横断面 |
| cryogenic [ˌkraiəu'dʒenik] a. | 低温学的；低温实验法的；制冷的,冷冻的 |
| cure [kjuə] vt. | 愈合,凝固 |
| cyclic ['saiklik] a. | 循环的 |

## D

| | |
|---|---|
| decade ['dekeid] n. | 十年 |
| decline [di'klain] v. | 降低 |
| defer [di'fə:] vt. | 拖延,推迟 |
| dense [dens] adj. | 密集的；稠密的 |
| deploy [di'plɔi] v. | 使用 |
| deposition [ˌdepə'ziʃən] n. | 沉积作用 |
| designate ['dezigneit] v. | 指定,标示 |
| desktop ['desktɔp] n. | [计算机]桌面 |
| detection [di'tekʃən] n. | 探测 |
| determine [di'tə:min] v. | 测定 |
| diagram ['daiəgræm] n. | 图解,图表 |
| diameter [daɪ'æmɪtə(r)] n. | 直径 |
| die [dai] n. | 模 |
| diffraction [di'frækʃən] n. | 衍射 |
| diffused [di'fju:zd] a. | 散布的,普及的,扩散的 |
| dimensional [di'menʃənəl] adj. | 空间的 |
| dispense [dis'pens] v. | 分发,分配 |
| dispersant [dis'pə:sənt] n. | 分散剂 |
| displacement [dis'pleismənt] n. | 排气量 |
| distort [dis'tɔ:t] vt. | 使变形 |
| distribution [ˌdistri'bju:ʃən] n. | 分布 |
| dominate ['dɔmineit] v. | 支配,占优势 |
| dowel ['dauəl] n. | 销 |
| drawbar ['drɔ:bɑ:] n. | 挂钩,拉杆 |
| drill [dril] n. | 钻床,钻孔机,钻子 |
| driveshaft ['draivʃɑ:ft] n. | 驱动杆；驱动轴,传动轴,主动轴 |
| dual ['dju(:)əl] adj. | 双重的 |
| ductile ['dʌktail] a. | 可延展的,有韧性的 |

durable ['djuərəbl] adj. 持久的,耐用的

# E

ease [iːz] n. 安乐,安逸,悠闲;v. 使……安乐,使……安心,减轻,放松
eject [i'dʒekt] vt. 顶出
elastic [i'læstik] adj. 有弹性的
elastomer [i'læstəmə(r)] n. 弹性体(弹胶物,合成橡胶,高弹体)
electrolytic [iˌlektrəu'litik] adj. 电解的
electrolytically [iˌlektrəu'litikəli] adv. 以电解
electron [i'lektrɔn] n. 电子
electroslag [i'lektrəuslæg] n. 电炉渣
embossing [im'bɔsiŋ] n. 浮雕(压纹,压花,模压加工)
encyclopedia [enˌsaiklǝu'piːdiǝ] n. 百科全书
engrave [in'greiv] v. 雕刻
espresso ['espresǝu] n. (蒸汽加压煮出的)浓咖啡
exclusively [ik'skluːsivli] ad. 仅仅,专门,唯一地
exterior [eks'tiǝriǝ] adj. 外部的
extraction [iks'trækʃǝn] n. 抽出,取出
extractive [iks'træktiv] adj. 抽取的
extruder [eks'truːdǝ] n. 挤出机,挤出设备
extrusion [eks'truːʃǝn] n. 挤出

# F

fabrication [ˌfæbri'keiʃǝn] n. 制作
facilitate [fǝ'siliteit] v. 帮助,使……容易,促进
feedscrew [fiːdskruː] n. 进给螺杆 螺旋给料机 螺旋送料机
ferrous ['ferǝs] a. 含铁的
filament ['filǝmǝnt] n. 细丝,细线,单纤维
filler ['filǝ] n. 填充料,掺入物,过滤器
filter ['filtǝ] n. 过滤器
fine [fain] a. 细微的
fitting ['fitiŋ] n. 日用器具
fixup ['fiksʌp] n. 修补
flaw [flɔː] n. 裂纹,瑕疵

flexibility [ˌfleksəˈbiliti] n. 柔韧性
flexible [ˈfleksəbl] adj. 灵活的
flux [flʌks] n. 流出
foaming [ˈfəumiŋ] n. 发泡成型,发泡
foil [fɔil] n. 箔,金属箔,薄金属片
footnote [ˈfutnəut] n. 脚注(给……作脚注)
forefront [ˈfɔːfrʌnt] n. 前列,前沿,最前方
forge [fɔːdʒ] n. 熔炉,铁工厂
forging [ˈfɔːdəiŋ] n. 锻件,锻造(法)
forming [ˈfɔːmiŋ] n. 成形,成型
forum [ˈfɔːrəm] n. 论坛
fracture [ˈfræktʃə] vt. & vi. 折断
frame [freim] n. 框,结构,骨架
friction [ˈfrikʃən] n. 摩擦
fume [fjuːm] n. 烟雾,气味
furnace [ˈfəːnis] n. 炉子,熔炉

# G

galvanized [ˈgælvənaizd] adj. 镀锌的
gate [geit] n. 浇口
gear [giə] n. 齿轮
generation [ˌdʒenəˈreiʃən] n. 代,时代,产生
generic [dʒiˈnerik] a. 一般的,普通的,共有的
geometric [dʒiəˈmetrik] adj. 几何学的
geometry [dʒiˈɔmitri] n. 几何(学)
given [ˈgivn] n. (推理过程中的)已知事物
grain [grein] n. 晶粒
granular [ˈgrænjələ(r)] adj. 由小粒而成的;粒状的
graphitization [ˌgræfitaiˈzeiʃən] n. 石墨化
gripper [ˈgripə] n. 夹子(抓爪器)
grind [graind] vt. 研磨
guillotine [ˈgilətiːn] n. 轧刀,裁切机,剪床;
gypsum [ˈdʒipsəm] n. 石膏

# H

hammer [ˈhæmə] n. 铁锤

| | |
|---|---|
| hardness ['hɑːdnis] n. | 硬度 |
| harmonic [hɑːˈmɔnik] a. | 调和的,音乐般的,和声的;n. 和音,调波 |
| headstock ['hedstɔk] n. | 主轴承,床头箱,主轴箱 |
| hobbyist ['hɔbiist] n. | 业余爱好者 |
| hog [hɔg;(US) hɔːg] n. | 肥猪 |
| hollow ['hɔləu] a. | 中空的、空心的 |
| horizontally [ˌhɔriˈzɔntli] adv. | 水平地 |
| housing ['hauziŋ] n. | 房屋(外壳,外套,外罩,住宅,卡箍,遮盖物) |
| hydraulic [haiˈdrɔːlik] a. | 水力的,水压的 |
| hydraulics ['haiˈdrɔːliks] n. | 液压 |

# I

| | |
|---|---|
| imperial [imˈpiəriəl] a. | 帝国的,英制的 |
| implantation [ˌimplɑːnˈteiʃən] n. | 灌输 |
| implement ['Implimənt] v. | 实施,执行 |
| impregnation [ˌimpregˈneiʃn] n. | 注入;浸渍 |
| inconel ['inkəunəl] n. | 铬镍铁合金,因康镍合金 |
| incorporate [inˈkɔːpəreit] vt. | 把⋯合并 |
| incorporate [inˈkɔːpəreit] a. | 合并的,公司组织的,具体化的;v. 合并,组成公司 |
| indexability ['indeksəbiliti] n. | 指望 |
| infrared ['infrəˈred] n. | 红外线 |
| infusible [inˈfjuːzəbl] a. | 不能熔化的;难以熔化的;耐热的; |
| inherent [inˈhiərənt] a. | 内在的,固有的 |
| initiative [iˈniʃətiv] n. | 倡议,主动性(权),主动精神 |
| injection [inˈdʒekʃən] n. | 注射 |
| insert [inˈsəːt] n. | 镶嵌物,镶件 |
| insoluble [inˈsɔljubl] a. | 不能溶解的 |
| insurmountable [ˌinsəˈmauntəbl] adj. | 不能克服的,难以对付的 |
| integral ['intigrəl] a. | 整体的 |
| integration [ˌintiˈgreiʃn] n. | 集成 |
| integrity [inˈtegriti] n. | 完整性 |
| interchangeability [ˌintə(ː)ˌtʃeindʒəˈbiliti] n. | 可交换性 |
| interface ['intə(ː)feis] n. | 界面,接触面 |
| interfaced ['intəːfeist] a. | 界面上的,界面的 |

# Vocabulary

| | |
|---|---|
| intermediate [ˌintəˈmiːdjət] a. | 中级的,中间的;n. 中间体,媒介物 |
| intermetallic [ˌintə(ː)miˈtælik] adj. | 金属间化合的 |
| interstitial [ˌintə(ː)ˈstiʃəl] adj. | 组织间隙的,间质的 |
| intricate [ˈintrikit] a. | 错综复杂的 |
| involute [ˈinvəluːt] n. | 渐伸线 |
| ion [ˈaiən] n. | 离子 |

## J

| | |
|---|---|
| jargon [ˈdʒɑːgən] n. | 行话 |
| jig [dʒig] n. | 带锤子的钓钩 |
| joining [ˈdʒɔiniŋ] n. | 连接 |

## K

| | |
|---|---|
| keyway [ˈkiːˌwei] n. | 键沟 |
| keynote [ˈkiːnəut] n. | 主旨,重点,基调 |
| kinetic [kaiˈnetik] adj. | 运动的 |

## L

| | |
|---|---|
| laminating [ˈlæmineitiŋ] n. | 层压法(层合法,分成薄层,卷成薄片) |
| laser [ˈleizə] n. | 激光 |
| lathe [leið] n. | 车床 |
| latte [lʌti] n. | 拿铁咖啡 |
| leadscrew [liːdskruː] n. | 导杆 |
| longitudinally [lɔndʒiˈtjuːdinəli] ad. | 纵向地 |
| lubricate [ˈluːbrikeit] v. | 润滑,涂油 |
| luster [ˈlʌstə] n. | 光泽 |

## M

| | |
|---|---|
| machine [məˈʃiːn] v. | 机加工、加工 |
| machinist [məˈʃiːnist] n. | 机械师 |
| macrostructure [ˌmækrəuˈstrʌktʃə] n. | 宏观结构 |
| magnesium [mægˈniːzjəm] n. | 镁 |
| magnetic [mægˈnetik] adj. | 磁的,有磁性的,有吸引力的 |
| mainstream [ˈmeinstriːm] n. | 主流 |
| malleable [ˈmæliəbl] a. | 可锻造的,可塑的,易改变的,有延展性的,韧性 |

| | |
|---|---|
| | 的 |
| manipulate [mə'nipjuleit] v. | 操纵,利用,假造 |
| marble ['mɑ:bl] n. | 大理石 |
| mass [mæs] n. | 大量,大批 |
| mastery ['mɑ:stəri] n. | 精通,熟练 |
| mechanism ['mekənizəm] n. | 机械,机构,结构,机制,原理 |
| medieval [ˌmedi'i:vəl] a. | 中古的,中世纪的 |
| melt [melt] n. | 熔化,熔化物,熔体 |
| vt. & vi. | (使)融化,(使)熔解 |
| metallography [ˌmetə'lɔgrəfi] n. | 金属组织学,金相学 |
| metallurgist [me'tælədʒist] n. | 冶金家,冶金学家 |
| metallurgy [me'tælədʒi] n. | 冶金;冶金学;冶金术 |
| metalworking ['metəlˌwɜ:kiŋ] n. | 金属加工 |
| meteoric [ˌmi:ti'ɔrik] adj. | 流星的,昙花一现的 |
| metrology [mi'trɔlədʒi] n. | 测量 |
| microstructure ['maikrəu'strʌktʃə] n. | 微观结构 |
| milling ['miliŋ] n. | 磨 |
| mobile ['məubail] a. | 移动的 |
| modest ['mɔdist] adj. | 适度的 |
| modulus ['mɔdjuləs] n. | 模数 |
| mold [məuld] n. | 模具,铸型 |
| molybdenum [mə'libdinəm] n. | 钼 |
| monitor ['mɔnitə] vt. | 检测 |
| mount [maunt] v. | 增长,装上,爬上,乘马,安装 |
| mounted ['mauntid] a. | 安在马上的,裱好的 |

# N

| | |
|---|---|
| naval ['neivəl] adj. | 海军的 |
| necessitate [ni'sesiteit] v. | 使……成为必需,需要 |
| negate [ni'geit] vt. | 否定 |
| neural ['njuərəl] adj. | 神经的 |
| niche [nitʃ] n. | 适当的位置 |
| nickel ['nikl] n. | 镍 |
| nitride ['naitraid] n. | 氮化物 v. 渗氮 |
| Nitrogen ['naitrədʒn] n. | 氮 |

# Vocabulary

nitrogen [ˈnaitrədʒən] n.　　氮气
nozzle [ˈnɔzl] n.　　喷嘴
numerical [njuː(ː)ˈmerikəl] adj.　　数字的
nut [nʌt] n.　　螺母
nylon [ˈnailən] n.　　尼龙

## O

obsolete [ˈɔbsəliːt] adj.　　过时的
offset [ˈɔːfset] n.　　抵消,支派,位移,偏移;v.弥补,抵消
onward [ˈɔnwəd] adj.　　向前的,前进的 adv. 向前,前进,在先
opacity [əuˈpæsiti] n.　　不透明
optimization [ˌɔptimaiˈzeiʃən] n.　　最佳化,优化
optimum [ˈɔptiməm] n.　　最适宜　adj.最适宜的
organically [ɔːˈgænikli] adv.　　有机地
originally [əˈridʒənəli] ad.　　本来,原来,最初,重要的
ornament [ˈɔːnəmənt] n.　　装饰品
ornamental? [ˌɔːnəˈmentl] a.　　装饰的
overcapacity [ˌəuvəkəˈpæsəti] n.　　生产能力过剩
overhang [ˈəuvəhæŋ] n.　　突出部分;悬垂部分
overexposure [ˈəuvəriksˈpəuʒə] n.　　过度暴露
overlap [ˈəuvəˈlæp] v. n.　　重叠,重复
oxidation [ɔksiˈdeiʃən] n.　　氧化

## P

painting [ˈpeɪntɪŋ] n.　　涂漆,着色
paleolithic [ˌpæliəuˈliθik] adj.　　旧石器时代的
pallet [ˈpælit] n.　　托盘
palletize [ˈpæliˌtaiz] vt.　　码垛堆集,货盘装运
palm [pɑːm] n.　　掌状物
pan [pæn] n.　　平底锅
parallel [ˈpærəlel] a.　　平行的; ad. 平行地;n. 平行线
patron [ˈpeitrən] n.　　资助人
pattern [ˈpætən] n.　　型芯,式样,模式 vt. 模仿,仿造 vi. 形成图案
penetration [peniˈtreiʃən] n.　　穿过,渗透,突破
periphery [pəˈrifəri] n.　　外围,圆周

| | |
|---|---|
| permanent ['pə:mənənt] adj. | 永久性的 |
| perpendicular [ˌpə:pən'dikjulə] adj. | 垂直的,直立的 |
| perpendicularly [ˌpə:pən'dikjuləli] ad. | 垂直(笔直,纵) |
| photogrammetry [ˌfoutə'græmItri:] n. | 照相测量法 |
| pick [pik] v. | 分拣 |
| piercing ['piəsiŋ] n. | 冲孔加工 |
| pillar ['pilə] n. | 柱子 |
| pin [pin] n. | 大头针,针,拴 |
| pitch [pitʃ] n. | 程度,投掷,音高,节距 |
| plasticating ['plæstikeitiŋ] n. | 塑化 |
| plate [pleit] n. | 模板 v. 镀(覆以金属板,熨平) |
| platform ['plætfɔ:m] n. | 平台,台,站台 |
| plating ['pleitiŋ] n. | 电镀,镀敷 |
| plotter ['plɔtə] n. | 绘图仪 |
| plummet ['plʌmit] vi. | 垂直落下 |
| poisonous ['pɔiznəs] adj. | 有毒的 |
| polish ['pɔliʃ] v. | 抛光 |
| polycarbonate [ˌpɔli'kɑ:bənit] n. | 聚碳酸酯 |
| polyester ['pɔliestə] n. | 聚酯纤维,涤纶 |
| polyethylene [ˌpɔli'eθili:n] n. | 聚乙烯 |
| polymer ['pɔlimə] n. | 聚合体 |
| polypropylene [ˌpɔli'prəupili:n] n. | 聚丙烯 |
| polystyrene [ˌpɔli'staiəri:n] n. | 聚苯乙烯 |
| polyurethane [ˌpɔli'juəriθein] n. | 聚氨酯 |
| porosity [pɔ:'rɔsiti] n. | 孔隙率；密集气孔 |
| potential [pə'tenʃ(ə)l] n. | 潜能,潜力,电压 |
| pouch [pautʃ] n. | 小袋 |
| pound [paund] vt. & vi. | 连续重击 |
| power ['pauə] n. | 强国,大国 |
| precaution [pri'kɔ:ʃən] n. | 预防措施 |
| precipitation [prisipi'teiʃən] n. | 沉淀 |
| precipitation hardening | 淀积硬化 |
| precision [pri'siʃən] n. | 精确,精密度 |
| presentation [ˌprezen'teiʃən] n. | 描述,表示 |
| principle ['prinsəpl] n. | 原理,机理 |

| | |
|---|---|
| productivity [ˌprɔdʌk'tiviti] n. | 生产率 |
| proficient [prə'fiʃənt] adj. | 精通的,熟练的 |
| propagate ['prɔpəgeit] v. | 繁殖,传播,传送 |
| prototype ['prəutətaip] n. | 原型 |
| provision [prə'viʒən] n. | 规定,条款;准备,食物,供应品 |
| pulley ['puli] n. | 皮带轮 |
| pump [pʌmp] n. | 抽水机 |
| punch [pʌntʃ] n. | 冲压机,冲头,冲孔,凸模冲头 vt.(用冲床)冲,冲孔 |

## Q

| | |
|---|---|
| quadrant ['kwɔdrənt] n. | 象限 |
| quenching ['kwentʃiŋ] n. | 淬火 |

## R

| | |
|---|---|
| rack [ræk] n. | 架,行李架,拷问台;齿条 |
| radius ['reidjəs] n. | 半径 |
| ratio ['reiʃiəu] n. | 比,比率 |
| raw [rɔː] adj. | 生的 |
| ray [rei] n. | 射线 |
| rebate ['riːbeit, ri'beit] n. | 槽,榫头 |
| reciprocate [ri'siprəkeit] vi. | 直线往复运动 |
| reciprocating [ri'siprəkeitiŋ] a. | 往复运动的,n. 往复 |
| recrystallization [riˌkristəlai'zeiʃən] n. | 再结晶 |
| refine [ri'fain] vt. | 使变得完善 |
| refining [ri'fainiŋ] n. | 精炼 |
| refractory [ri'fræktəri] adj. | 耐熔的 |
| related [rɪ'leɪtɪd] adj. | 相关的 |
| rank [ræŋk] v. | 排列,列为 |
| rectify ['rektɪfaɪ] v. | 改正 |
| replicate ['replɪkeɪt] v. | 复制 |
| repeatability [ripiːtə'biliti] n. | 可重复性 |
| reproducible [ˌriːprə'djuːsəbl] a. | 可再生的,可复写的,能繁殖的 |
| resemble [ri'zembl] vt. | 类似于 |
| resharpen ['riː'ʃɑːp(ə)n] vt. | 再次磨尖 |

| | |
|---|---|
| residue ['rezidjuː] n. | 余渣 |
| resin ['rezin] n. | 树脂(松香,树脂状沉淀物,树脂制品) |
| resistance [ri'zistəns] n. | 抵抗力,反抗,反抗行动;阻力,电阻;反对 |
| resolution [ˌrezə'luːʃn] n. | 分辨率 |
| restrain [ris'trein] v. | 抑制,阻止,束缚 |
| restrict [ris'trikt] vt. | 限制,约束 |
| retailer [riː'teilə] n. | 零售商 |
| rigid ['ridʒid] a. | 刚性的 |
| robust [rə'bʌst] a. | 强壮的,强健的,坚固的,结实的,耐用的 |
| rod [rɔd] n. | 杆,棒 |
| rolling ['rəuliŋ] n. | 轧制 |
| rotary ['rəutəri] a. | 旋转 |
| rough [rʌf] a. | 粗糙的 |
| routing ['ruːtiŋ] n. | 特形铣 |
| runner ['rʌnə(r)] n. | 流道 |
| rust [rʌst] n./v. | 生锈 |

## S

| | |
|---|---|
| screw [skruː] n. | 螺杆,螺丝钉 |
| sculptural ['skʌlptʃərəl] a. | 雕刻的,雕塑的 |
| sculpture ['skʌlptʃə] n. | 雕塑 |
| sector ['sektə(r)] n. | 部门,领域 |
| segment ['segmənt] n. | 部分;段,节 |
| servo ['səːvəu] n. | 伺服 |
| shank [ʃæŋk] n. | 胫,刀杆,模柄 |
| shielded ['ʃiːldid] adj. | 有屏蔽的 |
| shrink [ʃriŋk] vt. & vi. | 收缩 |
| shrinkage ['ʃriŋkidʒ] n. | 收缩,缩小,减低 |
| signal ['signəl] v. | 表示 |
| significant [sig'nifikənt] adj. | 有重大意义的 |
| simulation [ˌsimju'leiʃən] n. | 模拟,仿真 |
| sinter ['sintə] v. | 烧结 |
| sintering ['sintəriŋ] n. | 烧结 |
| sisal ['sisəl] n. | 剑麻,西沙尔麻 |
| slab [slæb] n. | 平板 |

# Vocabulary

slag [slæg] n. 熔渣
sleeve [sli:v] n. 套筒
slice [slais] n. 切片
slide [slaid] n. 滑轨
slot [slɔt] n. 狭缝
smith [smiθ] n. 铁匠,锻工
soldering ['sɔldəriŋ] n. 软钎焊
solid ['sɔlid] n. 固体
sophistication [səfisti'keiʃən] n. 复杂性
squeeze [skwi:z] vt. & vi. 挤压
stain [stein] vt. & vi. 使染色
stainless ['steinlis] adj. 不锈的
stamping ['stæmpiŋ] n. 冲压件(模锻,冲击制品)
standardization [ˌstændədai'zeiʃən] n. 标准化
stationary ['steiʃ(ə)nəri] adj. 不动的
stick [stik] vt. & vi. 粘贴,卡住
strategy ['strætədʒi] n. 战略,策略
streamlined ['stri:mlaind] adj. 流线型的
strength [streŋθ] n. 强度
strip [strip] n. 长条,条状
stripper ['stripə] n. 卸料板
styrene ['stairi:n] n. 苯乙烯
submerged [səb'mə:dʒd] adj. 水下的
substrate ['sʌbstreɪt] n. 基底
subtractive [səb'træktɪv] adj. 减去的
suction ['sʌkʃən] n. 吸,吸入
sulfide ['sʌlfaid] n. 硫化物
sulfur ['sʌlfə] n. 硫
superalloy [ˌsju:pə'æloi] n. 超耐热合金,超级合金,高温高强度合金
swaging ['sweidʒiŋ] n. 模锻
swarf [swɔ:f] n. 木屑
synonymous [si'nɔniməs] adj. 同义的
synthesize ['sɪnθəsaɪz] v. 合成;综合
synthetic [sin'θetik] a. 合成的,人造的 n. 人工制品

# T

| | |
|---|---|
| tailings ['teiliŋz] n. | 残渣,尾矿 |
| tailor ['teɪlər] v. | 专门定制 |
| tailorable ['teilərəbl] adj. | (衣料等)可裁制成衣的 |
| tailstock ['teilstɔk] n. | 尾架(顶尖座,滑轮活轴,托柄尾部) |
| tap [tæp] n. | 攻螺纹 |
| tarnish ['tɑ:niʃ] v. | 金属失去光泽;n. 无光泽 |
| tempering ['tempəriŋ] n. | 回火 |
| template ['templit] n. | 模板,样板 |
| temporary ['tempərəri] adj. | 一次性的 |
| thermal ['θə:məl] adj. | 热的,热量的 |
| thermoplastic [ˌθə:mə'plæstik] a. | 热塑性的;塑料热塑的 n. 热塑性塑料 |
| thermoset ['θə:məset] n. | 热固性塑料 |
| thermosetting [ˌθə:məu'setiŋ] a. | 热硬化性的,热固性的 |
| throttle ['θrɔtl] n. | 节气门 |
| tide [taid] n. | 潮流,趋势 |
| tilt [tilt] n. | 倾斜 |
| tip [tip] v. | 装顶端 |
| titanium [tai'teinjəm] n | 钛 |
| toggle ['tɔgl] n. | 拨动开关 |
| tolerance ['tɔlərəns] n. | 公差 |
| toolholder ['tu:lˌhəuldə(r)] n. | 刀柄 |
| tooling ['tu:liŋ] n. | 模具 |
| toolroom ['tu:lru:m] n. | 工具室(工具车间) |
| top [tɔp] v. | 达到顶端 |
| topography [tə'pɔgrəfi] n. | 地形,地势,地貌 |
| touch [tʌtʃ] v. | 粘连 |
| toughness ['tʌfnis] n. | 韧性 |
| track [træk] n. | 轨道 |
| tradeoff ['treidˌɔ:f] n. | 折中,权衡 |
| transfer [træns'fə:] v. | 传递 |
| transparency [træns'pɛərənsi] n. | 透明性 |
| transverse ['trænzvə:s] adj. | 横向的 |
| troublesome ['trʌblsəm] adj. | 令人烦恼的,讨厌的 |

# Vocabulary

trunnion ['trʌnjən] n.  枢轴
tube ['tjuːb] n.  管,软管
tumbler ['tʌmblə] n.  不倒翁
turbocharger ['təːbəuˌtʃɑːdʃə] n.  涡轮增压器
turret ['tʌrit] n.  小塔

## U

ubiquitous [juːˈbikwitəs] a.  无所不在的,普遍存在的
ultrasound [ˈʌltrəˌsaund] n.  超声波
ultraviolet [ˈʌltrəˈvaiəlit] adj.  紫外光的
undercut [ˈʌndəkʌt] n.  侧凹,侧抽芯
upset [ʌpˈset] vt. & vi.  镦粗

## V

vacuum [ˈvækjuəm] n.  真空
valve [vælv] n.  阀门
vapor [ˈveɪpə] n.  蒸汽
vaporized [ˈveipəraiz] a.  蒸汽的,雾状的
variation [ˌvɛəriˈeiʃən] n.  变化
versatile [ˈvəːsətail] adj.  多用途的
versatility [ˌvɜːsəˈtiləti] n.  通用性
version [ˈvəːʃən] n.  形式,种类
vertical [ˈvəːtikəl] adj.  垂直的
vessel [ˈvesl] n.  容器
virtually [ˈvɜːtjʊəli] adv.  几乎,实际上
visualization [ˌvizjʊəlaiˈzeiʃən] n.  可视化
volume [ˈvɔljuːm] n.  体积,容积
volumetric [vɔljuˈmetrik] adj.  测定体积的

## W

wafer [ˈweɪfə(r)] n.  晶片
washstand [ˈwɔʃstænd] n.  盥洗盆、盥洗台
wavy [ˈweivi] adj.  有波浪的(锯齿状)
way [wei] n.  滑道,(导)轨
welding [ˈweldiŋ] n.  焊接

| | |
|---|---|
| woodruff [wudrʌf] n. | 半圆 |
| workpiece ['wə:kpi:s] n. | 工件 |
| wrap [ræp] vt. | 包,裹 |

## Z

zinc [ziŋk] n.　　　　　　　锌